Urban Energy Systems

Scrivener Publishing
100 Cummings Center, Suite 541J
Beverly, MA 01915-6106

Publishers at Scrivener
Martin Scrivener (martin@scrivenerpublishing.com)
Phillip Carmical (pcarmical@scrivenerpublishing.com)

Urban Energy Systems

Modeling and Simulation for Smart Cities

Edited by
Deepak Kumar

WILEY

This edition first published 2023 by John Wiley & Sons, Inc., 111 River Street, Hoboken, NJ 07030, USA and Scrivener Publishing LLC, 100 Cummings Center, Suite 541J, Beverly, MA 01915, USA
© 2023 Scrivener Publishing LLC
For more information about Scrivener publications please visit www.scrivenerpublishing.com.

All rights reserved. No part of this publication may be reproduced, stored in a retrieval system, or transmitted, in any form or by any means, electronic, mechanical, photocopying, recording, or otherwise, except as permitted by law. Advice on how to obtain permission to reuse material from this title is available at http://www.wiley.com/go/permissions.

Wiley Global Headquarters
111 River Street, Hoboken, NJ 07030, USA

For details of our global editorial offices, customer services, and more information about Wiley products visit us at www.wiley.com.

Limit of Liability/Disclaimer of Warranty
While the publisher and authors have used their best efforts in preparing this work, they make no representations or warranties with respect to the accuracy or completeness of the contents of this work and specifically disclaim all warranties, including without limitation any implied warranties of merchantability or fitness for a particular purpose. No warranty may be created or extended by sales representatives, written sales materials, or promotional statements for this work. The fact that an organization, website, or product is referred to in this work as a citation and/or potential source of further information does not mean that the publisher and authors endorse the information or services the organization, website, or product may provide or recommendations it may make. This work is sold with the understanding that the publisher is not engaged in rendering professional services. The advice and strategies contained herein may not be suitable for your situation. You should consult with a specialist where appropriate. Neither the publisher nor authors shall be liable for any loss of profit or any other commercial damages, including but not limited to special, incidental, consequential, or other damages. Further, readers should be aware that websites listed in this work may have changed or disappeared between when this work was written and when it is read.

Library of Congress Cataloging-in-Publication Data

ISBN 9781119847441

Cover image: Networking Metropolis, AICANDY | Dreamstime.com
Cover design by Kris Hackerott

Set in size of 11pt and Minion Pro by Manila Typesetting Company, Makati, Philippines

Printed in the USA

10 9 8 7 6 5 4 3 2 1

Contents

Preface	xi
Acknowledgements	xix
List of Chapters and Affiliations	xxiii

1 Emerging Trends of Urban Energy Systems and Management   1
Deepak Kumar
   1.1 Introduction   2
   1.2 Research Motivation   4
   1.3 Stand-Alone and Minigrid-Connected Solar Energy Systems   6
   1.4 Conclusion   12
         References   13

2 Transitions in the Urban Energy Scenario and Approaches   19
Deepak Kumar
   2.1 Introduction   20
   2.2 Recent Transformation in Energy Sectors   22
   2.3 Research Progressions   24
   2.4 Breaking the Cycle   25
   2.5 Conclusion   27
   2.6 Future Implications   27
         References   28

3 Urban Renewable Energy Resource Optimization Systems   31
Kalpit Jain and Devendra Kumar Somwanshi
   3.1 Introduction   32
   3.2 Literature Review   33
        3.2.1 Long-Term Sustainable Solar Power Generation   33
              3.2.1.1 Common Issues of Long-Term Sustainable Solar Power Generation   39
              3.2.1.2 Strengths and Weakness Strength   40

	3.3	Conclusion	43
		References	44
4	**Approaches for District-Scale Urban Energy Quantification and Rooftop Solar Photovoltaic Energy Potential Assessment**		47
	Faiz Ahmed Chundeli and Adinarayanane Ramamurthy		
	4.1	Introduction	48
	4.2	District-Scale Urban Energy Modelling	49
		4.2.1 "Bottom-Up" Modelling Approach – Archetype	49
		4.2.2 The Renewable Energy Modelling Approach	50
		4.2.3 Urban Microclimate	50
	4.3	Evaluation of Energy Performance – The Case in Chennai	52
		4.3.1 Profile of the Case Area	52
		4.3.2 Data Model and Construction Techniques	53
		4.3.3 Archetype Classification	53
		4.3.4 Energy Quantification	55
		4.3.5 Analysis of the Archetype Energy Quantification	57
		4.3.6 Solar PV Potential Calculation	57
		4.3.7 Analysis of Solar PV Potential	58
		4.3.8 Scaling of Archetype Building Energy to District-Scale Urban Energy	58
		4.3.9 Scaling of Archetype PV Potential to District-Scale PV Potential	59
	4.4	Discussions and Conclusions	60
		4.4.1 Discussion	60
	4.5	Conclusions	61
		References	62
5	**Energy Consumption in Urban India: Usage and Ignorance**		65
	Rajnish Ratna and Vikas Chaudhary		
	5.1	Background	66
	5.2	Introduction	67
	5.3	Energy Outlook for India	68
	5.4	Power Demand and Resources in India	71
	5.5	Energy and Environment	73
	5.6	Sustainable Development Goals (SDGs) for Indian Electricity Sector	75
	5.7	Results	78
	5.8	Conclusions	78
		References	79

6 Solar Energy from the Urban Areas: A New Direction Towards Indian Power Sector 81
Sonal Jain

    6.1 Introduction 81
    6.2 Renewable Energy Chain in India 83
    6.3 Development of Solar Photovoltaic and Solar Thermal Plants 84
    6.4 Solar Photovoltaic Market in India 85
    6.5 Need for Solar Energy 86
    6.6 Government Initiatives 86
    6.7 Challenges for Solar Thermal Systems 87
    6.8 Benefits of Solar PV 88
    6.9 Causes of Delay in Solar PV Implementation and Ways to Quicken the Rate of Installation 89
    6.10 Future Trends of Solar PV 90
    6.11 Conclusion 90
    References 91
    Other Works Consulted 92

7 Energy Management Strategies of a Microgrid: Review, Challenges, Opportunities, Future Scope 93
Chiranjit Biswas, Somudeep Bhattacharjee, Uttara Das and Champa Nandi

    7.1 Introduction 93
    7.2 Methodology 95
        7.2.1 Research Studies Selection Criteria 95
        7.2.2 Section of Literature 95
        7.2.3 Testing Criteria 95
        7.2.4 Extraction of Data 96
        7.2.5 Findings 96
    7.3 Preliminary 97
        7.3.1 Fuzzy Logic–Based Management Strategies 97
        7.3.2 AI-Based Management Strategies 103
        7.3.3 Other Management Strategies 106
    7.4 Challenges of Energy Management in Microgrids 111
    7.5 Opportunities 112
    7.6 Future Research Direction 113
    7.7 Conclusion 113
    References 114

viii CONTENTS

8 Urban Solid Waste Management for Energy Generation 119
Shikha Patel and Reshmi Manikoth Kollarath
   8.1 Introduction 119
        8.1.1 Background 119
        8.1.2 Study Focus 121
   8.2 Literature Review 122
   8.3 Methodology 125
        8.3.1 Formulating Research Background 125
        8.3.2 Literature Review 126
        8.3.3 Analysis 127
   8.4 Case Study 127
        8.4.1 Precedent Success 127
        8.4.2 Precedent Failure 128
        8.4.3 The Takeaway from Case Studies 130
   8.5 Research Findings: Challenges of Waste-to-Energy Conversion 130
        8.5.1 Environmental Challenges 131
        8.5.2 Technological Challenges 132
        8.5.3 Social Challenges 132
        8.5.4 Economic Challenges 133
   8.6 Recommendations 134
   8.7 Conclusions and Discussion 135
        Acknowledgements 136
        References 136

9 Energy from Urban Waste: A Mysterious Opportunity for Energy Generation Potential 141
Shivangini Sharma and Ashutosh Tripathi
   9.1 Introduction 142
   9.2 Scenario of Solid Waste Management of Various Countries Around the World 143
   9.3 Waste-to-Energy Processes 147
   9.4 Challenges to Waste-to-Energy Generation 153
   9.5 Conclusion 154
        References 155

10 Sustainable Urban Planning and Sprawl Assessment Using Shannon's Entropy Model for Energy Management 157
Pranaya Diwate, Priyanka Patil, Pranali Kathe and Varun Narayan Mishra
   10.1 Introduction 158

10.2	Study Area	159
10.3	Materials and Methodology	160
	10.3.1 Satellite Data Used	160
	10.3.2 Pre-Processing of Satellite Data	160
	10.3.3 Accuracy Assessment	162
	10.3.4 LULC Change Detection	162
	10.3.5 Shannon Entropy Model	162
10.4	Results and Discussion	163
	10.4.1 LULC Maps	163
	10.4.2 Accuracy Assessment	163
	10.4.3 LULC Change Detection	165
10.5	Conclusion	168
	Acknowledgements	169
	References	169

11 Sustainable Natural Spaces for Microclimate Mitigation to Meet Future Urban Energy Challenges — 171
Richa Manocha and Deepak Kumar

11.1	Introduction	172
11.2	Nature and Human Connection	174
11.3	Urban Gardening	176
11.4	Urban Greening and Energy Benefits	177
11.5	Nurturing a Connection to Nature in Early Years	177
11.6	Conclusion	180
11.7	Future Implication	181
	References	181

12 Synthesis and Future Perspective — 193
Deepak Kumar

12.1	Introduction	193
12.2	Synthesis of the Research	195
12.3	Future Urban Energy Policies, and Initiatives	199
12.4	The Challenge Ahead	201
12.5	Strategies for Improvement	201
	References	203

About the Editor — 205

Index — 207

Preface

Urban Energy Systems refer to the systems that supply energy to urban areas for various purposes, such as electricity generation, heating, cooling, and transportation. These systems are becoming increasingly important due to the growth of urbanization and the need to reduce carbon emissions and increase energy efficiency. Some examples of urban energy systems include:

a) *Energy systems:* Aims to develop efficient and sustainable heating and cooling systems for urban areas. This includes research on the use of renewable energy sources including geothermal, solar thermal, and waste heat recovery.
b) *Energy-efficient buildings:* Focuses on developing building technologies and systems to reduce energy consumption and improve indoor air quality. It includes research on building envelope design, lighting systems, HVAC systems, and building automation.
c) *Electricity generation and distribution systems:* Includes power plants, transmission lines, and distribution networks to provide electricity to homes, businesses, and public facilities.
d) *Heating and cooling systems:* Includes heating and cooling networks which use central heating and cooling plants to supply energy to multiple buildings in a neighbourhood or region.
e) *Transportation systems:* Includes public transportation networks like buses, trains, and subways to provide energy-efficient alternatives to personal vehicles. The research aims to develop energy-efficient and sustainable transportation options, such as electric vehicles and public transportation systems to reduce emissions and congestion in urban areas.

f) *Renewable energy systems:* Include solar panels, wind turbines, and other renewable energy sources to be built an urban energy system to reduce reliance on fossil fuels towards a decrease in carbon emissions.
g) *Smart grid technology:* Focuses on developing advanced technology and control systems for electricity generation, transmission, and distribution. Smart grids incorporate renewable energy sources, energy storage, and demand response mechanisms to increase the efficiency and reliability of the electricity grid.
h) *Energy policy and planning:* Focuses on developing policies and strategies to promote sustainable and resilient urban energy systems, including incentives for renewable energy development, energy efficiency programs, and urban planning strategies to reduce energy consumption.

Research on urban energy systems is a rapidly evolving field, with a focus on developing sustainable and resilient energy systems for urban areas. Hence, the study on urban energy systems plays a critical role in ensuring the sustainability and resilience of urban areas, and there is increasing focus on developing innovative and efficient systems to link the energy needs of growing cities with developing sustainable and resilient cities with innovation and collaboration among researchers, industry, and policymakers to meet the energy requirements of urban areas.

The development of sustainable and smart cities requires the integration of various energy systems and technologies to meet the increasing demand for energy while minimizing environmental impacts. Modelling and simulation can provide valuable insights into the performance of urban energy systems and support decision-making for sustainable urban energy planning. A modelling and simulation perspective for sustainable smart cities involves developing models of energy systems and simulating their performance under different scenarios. These models can help urban planners and policymakers evaluate the feasibility and effectiveness of various energy technologies and strategies. In general, a modelling and simulation approach can be used to evaluate the performance of renewable energy sources, such as solar and wind power, in an urban context. The model can consider factors such as the location, orientation, and capacity of the renewable energy system and simulate its performance under different weather conditions and energy demand scenarios. Likewise, models can be developed to simulate the performance of district heating and cooling systems, electric vehicle charging infrastructure, and smart grid technologies.

These models can help evaluate the impact of different energy policies and incentives on the adoption of sustainable energy technologies and the overall performance of urban energy systems. Modelling and simulation can also be used to identify potential risks and vulnerabilities of urban energy systems, such as the impact of extreme weather events or cyber-attacks on the electricity grid. By simulating these scenarios, urban planners and policymakers can develop strategies to improve the resilience of urban energy systems and reduce the risk of disruptions to the energy supply.

In summary, a modelling and simulation perspective for sustainable smart cities can provide valuable insights into the performance and resilience of urban energy systems and support decision-making for sustainable urban energy planning. By integrating various energy systems and technologies, urban areas can achieve a more sustainable and resilient energy future. However, few studies have been reported using the harmonized approach of core science and research basics, as there are larger concerns about capacity building to use urban energy systems in achieving sustainable development goals. The book entitled "Urban Energy Systems with A Modelling and Simulation Perspective for Sustainable Smart Cities" contains chapters authored by reputed academicians, researchers, scientists, administrators, urban planners, professionals, policymakers, planners, and experts working in the field of urban planning and sustainable energy management. It will be useful for researchers, Post Graduate Students, Policy makers, Scientists, academicians, IoT researchers, professionals, data scientists, data analysts, practitioners, and people who are interested in the identification and development of methodologies, frameworks, tools, and applications through quantitative/qualitative results, and discussions.

Organization of the Book

The book is organized into twelve chapters. A brief description of each of the chapters follows:

Chapter 1 identifies the existing challenges in the management of urban energy resources in the new millennium. The chapter sets the section for discussions presented for the emerging trends of urban energy systems and management with a modelling and assessment perspective for sustainable smart cities. This incident catalogues the global orientation of energy and the related difficulties to identify the significance of forming energy policies, plans and frameworks. This chapter discusses Solar, hydro, biomass, geothermal, tidal, and wind as renewable energies to validate that renewable and low-carbon energy is clean energy. Energy models evaluate for an improved design, new policies, and related technologies for urban energy systems. There has been an explosion of models during the past few decades, covering a wide range of formulations, applications, periods, and geographic regions. This study is very useful for investigating the changes in energy scenarios due to human activity in an urban area.

Chapter 2 produces the need for transitions in the urban energy scenario and approaches for the development of policy. The chapter contends that transitions in excess renewable energy are the subject of research and can be used even in countries that don't have access to the advances in technology. Due to rising prices and dwindling oil supplies, energy efficiency measures are becoming more desirable. Hence, such resources have a major proposition in the whole energy portfolio as many of them are focusing on thrusting nature-friendly authority.

Chapter 3 takes logical alignment about the rights and wrongs in the necessity for a renewable energy source that won't harm the environment is growing. The creation of renewable energy comes from various sources like long-term sustainable Solar PV generation, optimization, solar resources, wind energy, hybrid solar-wind power generation, solar thermoelectric hybrid systems and microgrid systems. Several techniques are employed in various systems by diverse researchers in renewable sources of energy. The author examines some challenges in energy resources. The purpose of this chapter is to consider issues about approaches to thinking over various concerns.

Chapter 4 reviews the elements of energy usage that could be encouraged in methods for district-scale urban energy quantification and rooftop solar photovoltaic energy potential assessment. The authors argue that for energy-efficient urban design and planning, understanding the energy demand and consumption at the city scale is essential. This chapter uses a simplified methodology for the quantification of consumption and a possible supply of energy demands of an urban district by classifying them into archetypes, a bottom-up urban energy modelling technique. A study is conducted for detailed the weighted average net energy demands of the archetypes are then scaled to the selected study area using GIS. A district-scale 3D urban energy map and solar PV potential map of the case area is generated for simplified interpretations. The present study will be helpful for sustainable water resource management and agricultural applications.

Chapter 5 reviews the energy consumption in urban India about usage and its ignorance. Energy consumption and human-triggered carbon dioxide emissions rank India as the world's third-highest emitter. The current pattern of energy usage (particularly based on fossil-based fuels) raises severe concerns. There is no comprehensive baseline urban energy-use dataset for all urban districts in India that is comparable to national totals and that integrates social, economic, and infrastructure factors. This chapter focuses on urban energy, namely renewable and sustainable options.

Chapter 6 presents an analysis of issues related to solar energy from urban areas as a new direction towards the Indian power sector. It articulates that solar energy systems are now widely available for business and household usage and require very little maintenance. Solar power is increasingly popular in wealthier countries. This study is about the micro-level planning and development of energy resources available. The basic objectives of the study will be very useful for decision-makers and planners to prepare for global climate change.

Chapter 7 addresses the issue related to energy management strategies of a microgrid as a review, of the challenges, opportunities, and future scope. This chapter attempts to analyze some research papers to acquire about the limitations of microgrid energy management systems and discover energy in a microgrid in a much smarter way. This part provides essential background for our ongoing efforts to report the approaches for controlling the power in microgrids. The techniques employed in this study deliver valuable information to study some formulations for scheduling models through some intellect approaches to investigate some diverse models.

Chapter 8 analyses and compares recent approaches for development in waste management systems. The author systematically addresses the theoretical assumptions for urban solid waste management for energy generation. This paper explores the authoritative knowledge of existing waste management practices and policies in Bangalore, India. The aim is to relate two critical issues: managing waste in large cities and converting it into energy once managed. The methodology allows a thorough literature review investigating the existing data and with the help of surveys and non-structured interviews with experts, identifying challenges of managing waste and transforming it into energy. The findings wind up in three scales, namely, city, neighbourhood, and household scales. The analyzed data, shaped by an understanding of the challenges at these scales and drawn from the interviews and surveys' conclusions, is categorized into four themes: environmental, technological, social, and economic aspects. The results recommend policies to abate these challenges, promote growth, and foster a better transition towards more sustainable development.

Chapter 9 reviews issues related to the energy from the urban waste concept as a mysterious opportunity for energy generation potential. The authors argue that One of the most practical concepts for dealing with this waste is the consumption for gaining energy. Various advanced technologies such as methane capture technology, plasma-pyrolysis, gasification, incineration, bio-methanation, etc., can extract useful energy from this legacy waste. All these processes work on different components of municipal solid waste. This chapter talks about how the global, as well as local scenario of municipal solid waste management, has altered over time. Also, there is a discussion about the various processes and technologies which can yield energy from waste. In the last portion of the chapter, various gaps in the implementation of these process technologies have been discussed. It can be concluded that efficient solid waste management can be achieved by the application of "waste-to-energy" and "circular economy" concepts.

Chapter 10 discusses generic concepts of sustainable urban planning and sprawl assessment using Shannon's entropy model for energy management. In the current study, temporal remote sensing datasets of the years 1990, 2000, 2010, and 2020, along with secondary data, are used for detecting LULC changes in Gandhinagar district, India. The area under investigation saw an increase in built-up land at the expense of loss in the vegetated surface. The present study highlights the cost of using remote sensing approaches to examine the type and level of ongoing changes. The study area saw a quick increase in the number of buildings at the expense of the natural cover. This work also emphasizes the use of remote sensing pictures in creating efficient master plans and management for regulated urban expansion at both the regional and local levels.

Chapter 11 presents the idea of sustainable natural spaces for microclimate mitigation to meet future urban energy challenges and the role agent technology can play in security management. This research examines the literature on urban areas to draw links between humans and the environment, as well as the distance between them and strategies to bridge them. Urban greening can reestablish this link and reduce carbon footprint and energy use. Urban regeneration and urban planning continue to prioritize tools to stimulate expansion within pre-The chapter narrates about existing structures that are subject to climate change issues. existing urban districts, despite opposing demands.

Chapter 12 concludes with ideologies necessary for the synthesis of the research, future urban energy policies, and initiatives as a future perspective in the new millenniumThe study talks about the lack of available space for renewable energy installations as a major obstacle in urban areas. The equilibrium between city energy demand and renewable energy density serves as the basis for our analytical methodology for decarbonized urban environments. The energy requirements of modern cities while also lowering their carbon footprint, and widespread adoption of renewable energy sources are essential with Improvements in efficiency, usability, cost-effectiveness, accessibility, and sustainability.

Waves of innovation in the energy sector could come from a variety of sources in the future. These sources include solar electricity generated in space, nuclear power facilities that can be disassembled and reused, and deep geothermal systems. Greenhouse gas emissions and the demand for clean water and air may be reduced if these strategies are implemented. The current edited book entitled *"Urban Energy Systems"* comprises chapters written by prominent researchers and experts. The key focus of this edited book focuses on a modelling and simulation perspective for sustainable smart cities to replenish the available resources by integrating the concepts, theories, and experiences of experts and professionals in this field.

Dr. Deepak Kumar
State University of New York at Albany,
New York, Albany, USA

Acknowledgements

"When you want something, the entire Universe conspires in helping you to achieve it."

—*Paulo Coelho*

The journey of the research is a unique experience of my life. It was my first time to be so focused on one topic to achieve so much from it. Since childhood, I was curious about the subject of science and have always been eager to learn things. After the completion of PhD study, I have realized that the joy of exploration comes only from the varieties of challenged topics rather than routine research.

I would like to express our sincere thanks to all the contributors who generously shared their knowledge and expertise for this book. Without their hard work and dedication, this work would not have been possible. I am also grateful to my publishing team for their invaluable assistance in bringing this project to fruition. Their meticulous attention to detail and tireless efforts have ensured that the content is accurate, accessible, and engaging.

Dr. Sulochana Shekhar has been an astounding mentor and I am truly grateful for her time and patience. I would like to also thank her to inspire me in choosing research and teaching as my career options. She always extended her immense support, and guidance along with sharing scientific knowledge. She trained me to think in the most critical & independent ways and to gain confidence in various aspects of life. It was an honor to work with her and will remain a role model for me. Her strong sense of practicality and critical thinking helped me, particularly at times of "brain-freezing" events. She cared for me like her son. I enjoyed and will always remember the parental care and attention from Dr Shekhar (that's why I use to call her 'Mummy Ji'). I owe many thanks to her for the inspiration and motivation to move forward. I always look forward to working with her in future endeavors. Moreover, I had been fortunate to receive guidance from Prof. Syed Ashfaq Ahmed for introducing me to the world of remote sensing and GIS and giving me a real interest in science and technology. I am very grateful to him for his continuous support and encouragement. I am also indebted to Dr. R. Nijgunappa for motivating me to enter the world of research and to undertake the PhD research. He will be one of the best faculty members in memory forever.

I would like to express my gratitude to my family, friends, and colleagues for their unwavering support and encouragement. Their love and encouragement have sustained me through the ups and downs of the writing and editing process. I would like to thank my whole family for all their love and encouragement to support me with significant persistence. My mother's moral support has been indispensable for her emotional support. I wish to express my sincere gratitude to my beloved sister for supporting me to accept this editorial assignment.

I have no words to thank my concealed source of inspiration (Dr. Shiti), who always inspired me like a mountain whenever I had a difficult time. Special thanks to her for enlightening me with aspects of life and for being there in my good and bad times with eternal concern. Her continuous sustenance helped me to stay focused on my current assignment without any pain. I am lucky to have her on my journey and for listening to my words with patience and a smile.

This assignment could not have been completed without the great support and encouragement of a few people. Regrettably, I cannot acknowledge by name all my fantastic support, because the list would not fit in here, but I gratefully thank each one of them including my wonderful friends at Amity University Uttar Pradesh (India). It would be a long list to mention all the people that I am indebted to, but Dr. Maya Kumari, Dr. Sabyasachi Chattopadhyay, and Ms. Shampa Dhar are a special mention for helping me in every manner. I would like to apologize for any person I forget to mention.

I would like to thank all contributors for their outstanding contributions to this book. Their insights, feedback, and guidance have been invaluable throughout the editing process. Earnest appreciation is extended to the reviewers for their constructive comments to improve the quality of the chapters.

Finally, I pay my obeisance to the almighty for being so kind to me during this period of trials and tribulations, whose grace has sustained me to reach this level of life. It has been a long journey when many obstacles stood in the way; yet the almighty sort me through, turning every failure into successful and enjoyable moments.

Thank you all for your invaluable contributions to this book.

Deepak Kumar
State University of New York at Albany,
New York, Albany, USA
April 2023

List of Chapters and Affiliations

Emerging Trends of Urban Energy Systems and Management
Deepak Kumar[1,2]*
[1]*Center of Excellence in Weather & Climate Analytics, Atmospheric Sciences Research Center (ASRC), University at Albany (UAlbany), State University of New York (SUNY), Albany, New York, USA*
[2]*Amity Institute of Geoinformatics & Remote Sensing (AIGIRS), Amity University Uttar Pradesh (AUUP), Gautam Buddha Nagar, Uttar Pradesh, India*

Transitions in the Urban Energy Scenario and Approaches
Deepak Kumar[1,2]*
[1]*Center of Excellence in Weather & Climate Analytics, Atmospheric Sciences Research Center (ASRC), University at Albany (UAlbany), State University of New York (SUNY), Albany, New York, USA*
[2]*Amity Institute of Geoinformatics & Remote Sensing (AIGIRS), Amity University Uttar Pradesh (AUUP), Gautam Buddha Nagar, Uttar Pradesh, India*

Urban Renewable Energy Resource Optimization Systems
Kalpit Jain[1]* and Devendra Kumar Somwanshi[2]†
[1]*Department of Mechanical Engineering, Sangam University, Bhilwara, Rajasthan, India*
[2]*Department of Electronics and Communication Engineering, Poornima College of Engineering, Jaipur, Rajasthan, India*

Approaches for District-Scale Urban Energy Quantification and Rooftop Solar
Photovoltaic Energy Potential Assessment
Faiz Ahmed Chundeli[1*] and Adinarayanane Ramamurthy[2]
[1]Department of Architecture, School of Planning and Architecture, Vijayawada, Andhra Pradesh, India
[2]Department of Planning, School of Planning and Architecture, Vijayawada, Andhra Pradesh, India

Energy Consumption in Urban India: Usage and Ignorance
Rajnish Ratna[1*] and Vikas Chaudhary[2†]
[1]Gedu College of Business Studies, Royal University of Bhutan, Gedu, Bhutan
[2]Manav Rachna University, Faridabad, India

Solar Energy from the Urban Areas: A New Direction Towards Indian Power Sector
Sonal Jain*
School of Social, Financial and Human Sciences, Kalinga Institute of Industrial Technology (KIIT) University, Bhubaneswar, India

Energy Management Strategies of a Microgrid: Review, Challenges, Opportunities, Future Scope
Chiranjit Biswas, Somudeep Bhattacharjee, Uttara Das
and Champa Nandi*
Department of Electrical Engineering, Tripura University, Agartala, Tripura, India

Urban Solid Waste Management for Energy Generation
Shikha Patel[1*] and Reshmi Manikoth Kollarath[2†]
[1]Department of Architecture and Urban Planning, Qatar University, Doha, Qatar
[2]BMS College of Architecture, Bangalore, India

Energy from Urban Waste: A Mysterious Opportunity for Energy Generation Potential
Shivangini Sharma and Ashutosh Tripathi*
Amity Institute of Environmental Sciences, Amity University Uttar Pradesh (AUUP), Gautam Buddha Nagar, Uttar Pradesh, India

Sustainable Urban Planning and Sprawl Assessment Using Shannon's Entropy
Model for Energy Management
Pranaya Diwate[1], Priyanka Patil[2], Pranali Kathe[2] and Varun Narayan Mishra[3]*
[1]*University Department of Basic and Applied Sciences, MGM University, Aurangabad, India*
[2]*Centre for Climate Change and Water Research, Suresh Gyan Vihar University, Jaipur, India*
[3]*Amity Institute of Geoinformatics & Remote Sensing (AIGIRS), Amity University Uttar Pradesh (AUUP), Gautam Buddha Nagar, Uttar Pradesh, India*

Sustainable Natural Spaces for Microclimate Mitigation to Meet Future Urban Energy Challenges
Richa Manocha[1]* and Deepak Kumar[1,2]†
[1]*Amity School of Business, Amity University Uttar Pradesh (AUUP), Gautam Buddha Nagar, Uttar Pradesh, India*
[2]*Amity Institute of Geoinformatics & Remote Sensing (AIGIRS), Amity University Uttar Pradesh (AUUP), Gautam Buddha Nagar, Uttar Pradesh, India*

Synthesis and Future Perspective
Deepak Kumar[1,2]*
[1]*Center of Excellence in Weather & Climate Analytics, Atmospheric Sciences Research Center (ASRC), University at Albany (UAlbany), State University of New York (SUNY), Albany, New York, USA*
[2]*Amity Institute of Geoinformatics & Remote Sensing (AIGIRS), Amity University Uttar Pradesh (AUUP), Gautam Buddha Nagar, Uttar Pradesh, India*

1

Emerging Trends of Urban Energy Systems and Management

Deepak Kumar[1,2]

[1]*Center of Excellence in Weather & Climate Analytics, Atmospheric Sciences Research Center (ASRC), University at Albany (UAlbany), State University of New York (SUNY), Albany, New York, USA*
[2]*Amity Institute of Geoinformatics & Remote Sensing (AIGIRS), Amity University Uttar Pradesh (AUUP), Gautam Buddha Nagar, Uttar Pradesh, India*

Abstract

Energy sustainability is crucial for all human activities, societal growth, and civilization. It tries to introduce the sustainable energy techniques and technology. Energy sustainability is crucial towards achieving the sustainability and it provides low environmental and ecological impacts, sustainable energy resources with high efficiency. It includes the living standards, societal acceptance, and equity. The energy from sun in form of heat and light are used to create a wide range of energy systems without causing climate change and environmental harm in any form. Solar, hydro, biomass, geothermal, tidal, and wind are renewable energies with better energy efficiency. Daily energy waste must be reduced to decrease costs and conserve natural resources.

Keywords: Urban energy, energy consumption, energy efficiency, renewable, energy intensity

---

Email: deepakdeo2003@gmail.com; ORCID: http://orcid.org/0000-0003-4487-7755

Deepak Kumar (ed.) Urban Energy Systems: Modeling and Simulation for Smart Cities, (1–18)
© 2023 Scrivener Publishing LLC

1.1 Introduction

Rapid energy consumption growth is a global concern, yet most residential solutions are based on qualitative studies with limited numbers of users in the industrialized world [1–3]. Recent work examines urban India's energy consumption behaviours, motivations, challenges, and other issues [4–6]. Small study samples limit their generalizability. Tradition, spirituality, or morality did not influence conservation, contrary to earlier findings [7–9]. Participants aren't concerned about sharing energy statistics. Contrary to prior studies, participants were also interested in automated energy control systems [10–12]. Information-sharing, appliance-level consumption disaggregation, and accessible manual controls are design options for this cohort.

The provision of energy services generally requires extensive use of energy resources. Power generation, transportation, illumination, temperature control, industrial procedures (such as refining and manufacturing), and many more fall under this category. Obtaining energy sources, transforming them into usable forms, transferring, disseminating, storing, and ultimately putting that energy to use is all part of the energy life cycle, which is a lengthy and intricate process [5, 13, 14]. Energy is essential because it paves the way for a comfortable lifestyle and helps society progress. Energy is being used in most countries in a way that cannot be maintained indefinitely. All nations are included in this category (developing, industrialized, etc.). The term "energy sustainability" refers to the practice of managing energy resources across their entire useful life cycle in a way that bolsters multiple characteristics of long-term sustainability [15–18].

In recent decades, both the absolute number and the share of India's population living in urban areas have grown steadily. Both new cities and old ones are becoming increasingly crowded. The majority of those moving away from their rural homes are heading to the cities [19–22]. Affected people's habits, routines, and top priorities have shifted as a result of urbanization. Increases in disposable income and educational attainment have coincided with a decline in family size and the introduction of cutting-edge information and communication technology in the workplace and the home [23, 24]. The outcome of these shifts is a more energy-hungry urban populace. At the same time, there has been a major shift in the fuel mix used by metropolitan households. We report the findings from surveys of energy consumption in three Indian cities, conducted to better understand the current state of urban energy consumption and the variables that are currently shaping it [5, 25].

Therefore, energy sustainability is a broad notion that includes but is not limited to the use of sustainable energy resources. As for the latter, they involve things like living standards, social acceptability, and fairness. These are analyzed in connection to one another. Several illustrations and examples have been given to help demonstrate the advantages of promoting energy sustainability [26, 27]. The graphics also emphasize how difficult it can be to improve energy sustainability by highlighting the complexities of energy sustainability and the components that contribute to it. The benefits and difficulties are especially clear in the case of net-positive energy buildings. The findings and conclusions can be used to teach and inform about energy sustainability and can act as a catalyst to push individuals and societies in the direction of that goal [29, 30]. Using solar panels to generate power is good for humanity and could significantly cut down on fossil fuel use. As long as the sun shines, this method of harnessing energy is both sustainable and environmentally friendly. Solar power, which has both advantages and disadvantages, is becoming increasingly common in residential and commercial buildings in major cities across the world [30, 31]. In rural areas, where blackouts are more frequent, the Indian government promotes the use of solar panels. These days, it's not uncommon to see solar-powered streetlights, especially in India's major cities. While solar is the most practical and widely available renewable energy source, there are numerous others. Increased electrical energy production from photovoltaic cells (also known as solar cells) may be possible in the not-too-distant future when more improved semiconductors are discovered. An energy source is said to be sustainable if it "meets the demands of the present without jeopardizing the ability of future generations to satisfy their own needs," a tenet of sustainability. Finding renewable energy sources, as opposed to finite ones, is central to the concept of sustainable power [32–34].

Since the national program does not call for a dramatic increase in the amount of commercial energy supplied to rural areas, the solution to rural energy shortages must rely on the efficient application of natural renewable energy sources like biomass, solar power, wind power, geothermal power, mini-hydro power, and the use of fuel-saving cookstoves. It was determined that the biogas power station was effective.

While traditional fuels have not been rapidly replaced in rural areas, they have been rapidly supplanted in metropolitan areas. Biomass is still used extensively for domestic energy production in rural areas. However, commercial fossil-based energy sources and electricity use are increasingly dominating the urban household energy mix. Despite this shift, the usage of biomass is still commonplace in modern urban dwellings. It has been noted that overall home energy usage in rural areas consistently exceeds that in urban areas. This is because people in cities are switching to more efficient fuel sources, while people in the countryside are still heavily reliant on less efficient solid fuels [35–37]. The energy consumption patterns of rural and urban households are distinct, with the latter favouring more cutting-edge fuels and services despite spending the same amount of money. There is a demand for higher-density fuel and electricity distribution because of the greater population density in urban areas and the corresponding decrease in available space for fuel storage and collection. In addition, utilities like electricity and fuel supply can be provided more cheaply in urban centres than in more remote, less densely populated places with limited purchasing power [38, 39]. When comparing rural and urban locations, the quality of energy services such as electricity is lower in the former. Due to these obstacles, improving rural residents' access to modern energy services is less likely to occur.

1.2 Research Motivation

Every day, the sun gives forth free energy that might be used to power numerous devices, yet this source of clean energy is often wasted. Similar to other forms of renewable energy, this precious resource cannot be saved in its raw form for later use. Therefore, one of the most practical and efficient uses is to transform it into energy and store the extra amount for later-use systems for capturing solar power. PV (Photovoltaic) systems, which convert sunlight directly into electricity, can be used to facilitate this change. The many benefits of solar energy have attracted interest from both developing and industrialized nations. The main benefits are their adaptability to both urban and rural settings, as well as their ease of exploitation, abundance, and reduced imposed costs compared to alternative means. Solar power generation sites with high capacity have been built in suitable regions around the world, but domestic-scale ones have not progressed sufficiently, especially in nonindustrialized countries. The problem is exacerbated by a lack of thorough investigation and scientific proof. Accordingly, the aforementioned argument can support the concept of researching solar power generation at home, especially during development. However, delivering constant load in isolated regions is complicated by the fact that solar systems are weather-dependent and subject to fluctuations in solar radiation.

Therefore, a PV system combined with an energy storage unit is a good option for off-grid properties. The battery is currently one of the most widely used types of energy storage. To meet the load demand in these far-flung places at a reasonable price and with a high degree of reliability, it is crucial to specify the ideal combination of power scheme components. Because of this, effective modelling and a robust optimization approach to addressing these issues are crucial. Mathematical modelling, optimal size, and techno-economic analysis of solar-based hybrid energy schemes have all been the subject of numerous published works. Using a genetic algorithm, this method optimizes the performance of microgrids powered by renewable energy sources like solar, wind, batteries, biodiesel, and hydrogen in the city of Tucson, Arizona, in the United States. The solar photovoltaic panel, battery storage units, inverter/converter system, and various other equipment and cables make up the stand-alone hybrid energy systems. Solar panels are the primary source of energy in this setup, and their output is sufficient to meet the need. When the sun isn't shining, the energy produced by PV panels is used to meet the load requirement, and any excess is stored in a battery bank for later use. The overcharged battery is then discarded. All of the system's components must be exclusively modelled, and then their optimal sizing must be determined based on the load requirement before the scheme can be optimized. Coal and oil-based fossil fuel generation poses a serious threat to the climate since it rapidly increases carbon emissions. The instability of oil prices over the past decade has led to dramatic growth in developing countries' capability to install solar photovoltaic (PV) panels.

While solar photovoltaic (SPV) generators have many uses due to their many advantages, their implementations can vary widely. The SPV-isolated systems are low-cost, risk-free, and easy-to-implement options for supplying power in a distributed manner. They make it possible to set up reliable, distributed power sources in remote places that are far from existing grids. Various industries, including telecommunications, agriculture, rural electricity, street lighting, signs, control, and rural development, make use of stand-alone PV systems. The water pumping system is a crucial use of PV systems in agriculture. Using this method, water can be withdrawn from remote locations where it would be prohibitively expensive to run a mains-powered connection.

MPP tracking and bidirectional power flow between the grid and PV are presented for a grid-connected motor-driven solar-powered water pumping system with efficient control. Off-grid microgrids, small electricity systems that may function without connection to a larger power grid, may play a crucial role in the evolution of decentralized renewable energy (RE) power grids. It is possible to generate enough power in nations where the national demand exceeds the normal production thanks to these networks, which are more cost-effective than extending transmission lines to remote areas. The lack of electricity, for instance, has a major negative impact on development in East Africa. Costly initial investments in large grids can impede the construction of infrastructure in underdeveloped nations. For rural communities in developing countries to take advantage of specific renewable energy sources, community-based microgrids are seen as the ideal alternative. "Off-grid users" are those who do not receive their electricity from a public or private utility. The authors define "off-grid" as a system and manner of life that does not rely on any external infrastructure, such as a central power plant. It's a way to plug into the grid in places with sparse populations and few power plants. This is synonymous with going off the grid or deciding to forgo the use of any public utilities. People that choose not to connect to the utility company's power grid employ alternative energy sources, such as solar panels, wind turbines, or a combination of the two, known as a "microgrid" or "minigrid," which are designed to serve relatively small populations. Potential renewable energy sources and methods to guarantee universal access to electricity for off-grid users were modelled, simulated, and optimized using the HOMER software (HOMER Pro, version 3.13.1). Thanks to its patented derivative-free technique, HOMER includes an integrated optimizer.

But lifting millions of Indians out of poverty at once is the actual problem facing the country. Heat waves and other forms of extreme weather are becoming increasingly common as a result of climate change in the country. India's main focus has been on making sure its decarbonization efforts are in line with SDG targets [40–43]. India must move considerably more quickly than other developed countries to build clean, inexpensive, and efficient energy systems if it is to provide access to energy and mobility, clean cooking fuel, food security, cooling, and catastrophe resilience for its citizens.

1.3 Stand-Alone and Minigrid-Connected Solar Energy Systems

Renewable energy sources (RES) are continuously gaining greater attention around the world as a solution to the growing problems of climate change, pollution, resource depletion, and energy consumption [36, 44, 45]. There is a growing and unavoidable demand for renewable energy to power the design and construction of microgrids and minigrids.

Comparing the cost of renewable energy to that of conventional energy, the latter comes out significantly cheaper. Energy issues for dispersed communities can be resolved by constructing freestanding minigrids, which use an unsustainable energy system. Supportable energy development and clean energy require substantial planning in developing countries with frequent power outages because of the economic implications. Therefore, the HOMER (hybrid optimization model of electric renewable) Pro program can build, prepare, and run the model in many settings, such as constrained and unconstrained systems, stand-alone, grid, and/or storage. Microgrids benefit from the advantages of the system design that allow for effective source loading and make life easier for power system operators. The value of HOMER is in the tools it provides for creating and simulating microgrid models [46, 47].

Scientists are working on ways to generate more power without negatively hurting the environment and at a lower cost. The lack of energy resources makes it more difficult for governments and citizens to plan for the future and achieve their development goals. Off-grid PV power plants were created and simulated using HOMER software to supply a rural county with electricity [48, 49]. Based on simulation results, an isolated PV system for a single-family home is the best option for providing electricity in off-grid locations. When comparing the system to a microgrid PV system that provides power to a rural town, the savings are substantial. Consistent with the foregoing, stand-alone PV power systems may be the best option for bringing electricity to rural areas. Such systems may also aid the country's government and environmental agencies in their quest to reduce the impact of natural disasters and encourage the growth of green energy sources works to meet its commitment to providing its citizens with access to clean, affordable power. It is also anticipated that the suggested stand-alone solar PV system will make an equal contribution to the future development of renewable energy generation systems in other nations with similar environmental, climatic, weather, and meteorological circumstances. Specifically, several countries in the region are predicted to make ideal candidates for such a system. Recently, there has been a global shift from fossil fuels to renewable energy sources due to rising electricity prices and greenhouse gas emissions (RES). It's possible to find the nicest weather in the world in a particular location. Due to a lack of available conventional hydrocarbons and rising GHG emissions from fuel combustion, the kingdom has ratified the Kyoto Protocol to lend international support to measures to slow the rate at which the planet warms. Electricity generation accounts for roughly more than half of all greenhouse gas emissions because of its heavy reliance on fossil fuels. Integration of RE and improvement of energy efficiency across sectors is central to the strategy proposed here for resolving this pressing problem [48, 50, 51].

Due to increased urbanization, the residential sector is viewed as a significant consumer of energy. One of the biggest issues with renewable energy systems is that the amount of power generated by PV and WTs can rise and fall depending on the weather. That being said, the most difficult challenge for any power-producing system is maintaining a constant rate of supply that is in sync with consumer demand. Therefore, the current difficulty is to guarantee the demand while managing the energy flow by regulating the produced power to supply the load and the surplus power to or from the battery/grid. To maximize efficiency and cut down on costs worldwide, several different approaches, control methods, and computer programs have been implemented. When it comes to home RE systems connected to the grid, the ideal approach is to factor in power rates, which have a major impact on the system's economic performance. ToU and step-rate tariffs are widely used in the electrical industry because of the positive impact they have on the grid's energy efficiency [52, 53]. Improved power grid reliability can be achieved by the use of solar and wind generation in tandem. To cut down on electricity costs and make money off the surplus, the economic power generation system relies heavily on home grid-tied PV/battery power. The proposed model was shown to be cost-effective through simulation. In addition, the Levelized cost of energy (LCOE) and the net present cost (NPC) are used in the economic analysis, both of which were determined with the help of the HOMER (Hybrid Optimization Model for Electric Renewable) program. It has been shown by Adeli *et al.* that PV is not capable of creating a net-zero energy building, hence alternative renewable energy sources should be utilized. It was determined that WT is a suitable replacement for PV generators in reducing electricity generation during the winter months. Furthermore, adjustment of the factors of reduced energy usage and the hours of resident thermal dissatisfaction made for a better condition for the building under study. To meet the needs of homes, engineers have developed a photovoltaic and energy storage system (ESS) that is connected to the grid. Keeping our energy sources secure is crucial to human progress and the evolution of our civilizations and civilization as a whole. This article provides a definition and analysis of energy sustainability, along with a look at some approaches and tools that can improve the situation. The significance and necessity of energy sustainability as a component of sustainability becomes apparent.

Low environmental and ecological impacts, sustainable energy resources and complementing energy carriers, high efficiency, and other characteristics are cited as necessities for improving energy sustainability.

Considerations such as quality of life, social acceptability, and fairness fall under the latter category, which is largely non-technical. It is hoped that the findings and conclusions will serve to inform and educate about energy sustainability and serve as a catalyst for increased energy sustainability. Throughout history, energy has been the driving force behind all human endeavours. Energy is essential for the existence of all living things in this biosphere. The energy from the sun affects everything, whether immediately or indirectly. The warmth and light of the sun are essential to the survival of all living things. The process of photosynthesis cannot occur without it. Over millions of years, depending on the type of fossil, the amount of pressure and heat, the remains of these organisms become fossilized, at which point they are used to create a wide range of fossil fuels. These fossil fuels are non-renewable because they have a finite supply and their creation takes a long time compared to the rate at which they are used in daily life. Greenhouse gases are released as a result of the combustion of fossil fuels including coal, crude oil, petroleum, and natural gas, which in turn causes climate change and environmental degradation.

Renewable energy is generated from resources that are naturally regenerated on a human timescale, such as the sun, water, plants, geothermal heat, wind, and tides. Renewable energy is often called "clean energy" because it produces no harmful emissions. Reduced energy consumption is fast becoming humanity's top priority. Every day, we lose a great deal of energy that should be preserved not only to reduce expenses but also to make better use of our natural resources over the long term.

As the availability of fossil fuels and other non-renewable resources decreases, people around the world are looking for alternative energy sources. To this end, photovoltaic systems, which convert solar energy into electricity, present a viable option. Such a system is eco-friendly since it generates no waste or toxins, and it may be used repeatedly. The need for energy in rural regions is increasing, and model approaches for a renewable energy supply have been created and proved to meet this need while also increasing economic productivity and contributing to a sustained improvement in rural residents' standard of living. This helps to cut down on carbon emissions by concentrating on technology that doesn't rely on fossil fuels, and also provides inputs for future rural energy interventions. Bringing power to India's rural communities is a complicated task. About 500 million people still live in rural areas without access to modern energy services; fuel quality is generally low, and energy is not used properly; the power supply is unstable, and access is restricted. Besides harming economic output, this also has serious repercussions for people's quality of life and the natural world. Degradation of the environment occurs on multiple scales due to the excessive use of locally sourced biomass and the rising reliance on fossil fuels (greenhouse gas – GHG emissions contributing to climate change). While lowering GHG emissions and local pollution, locally based solutions that use renewable energies to safeguard the rural power supply are creating new prospects for economic growth. The goal of the Rural Energy Supply Model (RESM) is to serve as a reliable resource for policymakers, entrepreneurs, specialists, and financiers.

The purpose is to fill in the gaps in our understanding of effective approaches for delivering power to remote places. This resource might be helpful in the planning of renewable energy projects to provide electricity to rural areas since it presents best practices in a coherent framework. RESM compiles the features, model-specific benefits, challenges, and success factors of various Rural Energy Supply models, with examples. There has been a dramatic shift in the global landscape of energy generation during the past decade due to the explosive growth of renewable generation. In the twenty-first century, there are millions of endpoints at which power is used. In India, the electrical grid has crumpled roughly over the past decade and the one before that. With the depletion of fossil fuels and other non-renewable resources, the search for new energy sources has become an urgent global priority. As a sustainable energy source that produces significantly less pollution than solar panels, photovoltaic energy provides hope in this quest. The planned research will centre on Economic Assessment models, which provide estimates of the local and state-level economic implications of developing and operating power generation. The work takes into account numerous facets of the problem, such as energy and economic ones, which is crucial for gauging the true results of investments. The report's findings help determine whether or not installing and maintaining solar systems is financially viable. Additional economic considerations involve things like conducting a cash flow analysis, estimating the energy cover factor with the results of the economic analysis, and conducting a sensitivity analysis for the most important physical and economic parameters; evaluating the costs of the PV systems (investment costs and costs for maintenance, servicing, and insurance against damage); and benefits due to gains for the avoided bill costs, the incentives, and the sold electricity; and so on. Since fossil fuels are running out everywhere, finding new sources of energy to power the modern world is essential.

The sun provides a renewable energy source that is clean, limitless, environmentally friendly, and potentially useful. The amount of electricity generated by photovoltaic (PV) systems is largely determined by the amount of solar radiation available, which in turn is affected by factors such as latitude, altitude, cloud cover, the quality of the local environment, the effectiveness of the PV technology, and societal and economic considerations. This research aims to establish criteria for evaluating the solar potential of rooftops in metropolitan areas. Using high-resolution satellite imagery (orthophotos), much geographic information system (GIS) technologies were utilized to evaluate the potential for urban solar power in a specific area. Extraction of rooftops from orthophotos, assessment of acceptable rooftop area for PV deployment, peak kW capacity estimation per building rooftop and determination of hourly solar PV energy generation potential are the three key components of the methodology. The photovoltaic (PV) potential is then calculated interactively in a raster format using proven models and algorithms for estimating the incoming sun brightness. The amount of sunlight reaching a roof from a variety of angles and orientations in different environments is calculated. The approach factors in the impacts of the weather, the site's latitude and elevation, the slope and aspect of the terrain, and the shadows cast by nearby buildings, chimneys, dormers, plants, and topography. In the future, researchers will need to collect data using a pyranometer to verify the method's efficacy and make the resulting data publicly available via a prototype.

The ability of humans to convert one form of energy to another has been a driving force in the rapid progress of our culture. Some energy is lost during this process. One way to improve energy efficiency is to replace inefficient machinery with more effective models. The use of energy sources has continued to rise indefinitely. Increasing urbanization, unsustainable development, industrialisation, transportation, industry, and many other factors are to blame for this upward trend. Consequences such as climate change, loss of biodiversity, abnormally quick melting of glaciers, depletion of the ozone layer, and harmful effects on public health are already becoming apparent as a result of the world's rising demand for energy. Traditional methods of transportation, lighting, heating, and more have all made use of renewable energy in the past. The wind was harnessed for usage by sailboats on the waves. Windmills were used to mill grains using the energy from the wind. There were several uses for biomass in the past. Over several centuries, however, people have favoured more inexpensive but dirtier energy sources like coal.

The usage of renewable energy, from residential rooftop solar panels to industrial-scale wind farms, is growing as new, cheaper methods of storing solar and wind power become available. Whole rural villages may rely solely on them to provide all of their energy needs, including those for cooking, heating, and lighting. Non-renewable energy sources are far more detrimental to the environment than renewable energy sources. Environmental implications are also substantial with renewable energy sources including geothermal, hydropower, and biomass. The technology employed, the location and a host of other variables all contribute to the kind and severity of environmental repercussions. To use renewable energy sources responsibly, it is important to have a firm grasp on the nature and extent of the environmental repercussions that doing so entails.

To reduce the daily operational cost of the residential grid-tied PV-WT battery system, this study provides an energy management method. To test the efficiency of the hybrid system and validate the energy balance and management approach, the system is run with increasing levels of self-consumption and Time of Use. The strategy is measured against a baseline approach in which it is applied without taking into account the operation cost or degradation of the BESS. There is a technical indicator analysis, a battery impact assessment, and an ecology investigation.

The influence of battery capacity change was analyzed to inform the comparison. In sum, it is concluded that there will be benefits in terms of both economics and the environment from investing in the installation and consumption of producing power, although it appears that starting domestic solar-based hydrogen production necessitates some backing plans from the government, such as incentives for importing equipment necessary for small-scale hydrogen-producing plants. It is crucial to evaluate the potential of using renewable energies for residential applications in light of the urgent need to decarbonize the energy sector. This makes it simple to install solar energy systems on the roofs of people's homes in the right places. This research set out to propose and evaluate a solar power generating system that might be used to generate the necessary electricity or hydrogen for residential use.

1.4 Conclusion

Cities are an immediate problem and opportunity for climate policy. Cities need the highest energy services because of high population density and economic activity concentration (urban energy demand). It's important to explain cities' significance in the world's energy systems and their connection to climate change. With more than half of the world's population living in cities and that number anticipated to climb to 75% by 2050, sustainable development must pass via cities. Growing cities bring problems to the environment, including energy as a natural resource, and quality of life.

Most cities understand the necessity of sustainability, both on a local and global basis. It demands viewing a city as a complex and dynamic ecosystem, open system, or cluster of systems, where energy and other natural resources are changed to meet urban needs. The key findings from the techno-economic analysis are as follows: If the proposed solar power plant is installed on the roof of a typical home, the homeowner could generate electricity as the surplus electricity in addition to meeting the electric power requirements of the home, earn good funds annually by selling the excess electricity to the grid, and reduce annual emissions of carbon dioxide by using the system in the first scenario.

Better knowledge and forecasting of power demand and power generation, especially in urban areas where the network is dense, are essential to the implementation of related policies currently underway in many nations throughout the world. Models of urban energy systems are essential for evaluating improved designs, new policies, and related technologies, as evidenced by the vast and varied body of material presented here. There has been an explosion of models during the past few decades, covering a wide range of formulations, applications, periods, and geographic regions. However, no work exists that brings to light the whole scale of the activity in this area, and thus no resource that helps interpret and make sense of this large body of writing. In this study, we aim to close that knowledge gap by conducting a systematic literature assessment on urban energy systems modelling.

References

1. P. J. Vergragt, L. Dendler, M. de Jong, and K. Matus, "Transitions to sustainable consumption and production in cities," *J. Clean. Prod.*, vol. 134, no. Part A, pp. 1–12, 2016, doi: 10.1016/j.jclepro.2016.05.050.
2. V. Janković, "A historical review of urban climatology and the atmospheres of the industrialized world," *Wiley Interdiscip. Rev. Clim. Chang.*, vol. 4, no. 6, pp. 539–553, 2013, doi: 10.1002/wcc.244.
3. V. Ganesh, S. Senthilmurugan, V. M. Ajay Krishna, R. R. Ajit Ram, and A. Prabhu, "Smart grid-meters and communications-design, challenges, issues, oppurtunities and applications," 2020, doi: 10.1109/ICADEE51157.2020.9368903.
4. G. Aiello, I. Giovino, M. Vallone, P. Catania, and A. Argento, "A decision support system based on multisensor data fusion for sustainable greenhouse management," *J. Clean. Prod.*, vol. 172, pp. 4057–4065, 2018, doi: 10.1016/j.jclepro.2017.02.197.
5. N. Suthar, *Estimating Effect of Electric Vehicles on Indian Grid System and Identifying Consequent Opportunities*, vol. 580. 2020.
6. K. Degirmenci, K. C. Desouza, W. Fieuw, R. T. Watson, and T. Yigitcanlar, "Understanding policy and technology responses in mitigating urban heat

islands: A literature review and directions for future research," *Sustain. Cities Soc.*, vol. 70, 2021, doi: 10.1016/j.scs.2021.102873.
7. L. J. Potgieter et al., "Managing Urban Plant Invasions: a Multi-Criteria Prioritization Approach," *Environ. Manage.*, vol. 62, no. 6, pp. 1168–1185, 2018, doi: 10.1007/s00267-018-1088-4.
8. E. Barrios-Crespo, S. Torres-Ortega, and P. Díaz-Simal, "Developing a dynamic model for assessing green infrastructure investments in urban areas," *Int. J. Environ. Res. Public Health*, vol. 18, no. 20, 2021, doi: 10.3390/ijerph182010994.
9. Q. Ding, L. Wang, M. Fu, and N. Huang, "An integrated system for rapid assessment of ecological quality based on remote sensing data," *Environ. Sci. Pollut. Res.*, vol. 27, no. 26, pp. 32779–32795, 2020, doi: 10.1007/s11356-020-09424-6.
10. I. Froiz-Míguez et al., "Design, implementation, and empirical validation of an IoT smart irrigation system for fog computing applications based on Lora and Lorawan sensor nodes," *Sensors (Switzerland)*, vol. 20, no. 23, pp. 1–33, 2020, doi: 10.3390/s20236865.
11. K. Kamau-Devers, V. R. Yanez, V. W. Medina Peralta, and S. A. Miller, "Using internal micro-scale architectures from additive manufacturing to increase material efficiency," *J. Clean. Prod.*, vol. 291, 2021, doi: 10.1016/j.jclepro.2021.125799.
12. D. Cannizzaro, A. Aliberti, L. Bottaccioli, E. Macii, A. Acquaviva, and E. Patti, "Solar radiation forecasting based on convolutional neural network and ensemble learning," *Expert Syst. Appl.*, vol. 181, 2021, doi: 10.1016/j.eswa.2021.115167.
13. U. I. Services, "Guide for Urban Integrated Hydro-Meteorological, Climate and Environmental Services Part 1a: Concept and Methodology," pp. 1–51, 2018.
14. L. Von Der Tann, N. Metje, H. Admiraal, and B. Collins, "The hidden role of the subsurface for cities," *Proc. Inst. Civ. Eng. Civ. Eng.*, vol. 171, no. 6, pp. 31–37, 2018, doi: 10.1680/jcien.17.00028.
15. S. Kumar, D. Sharma, S. Rao, W. M. Lim, and S. K. Mangla, "Past, present, and future of sustainable finance: insights from big data analytics through machine learning of scholarly research," *Ann. Oper. Res.*, 2022, doi: 10.1007/s10479-021-04410-8.
16. N. M. P. Bocken, S. W. Short, P. Rana, and S. Evans, "A literature and practice review to develop sustainable business model archetypes," *Journal of Cleaner Production*. 2014, doi: 10.1016/j.jclepro.2013.11.039.
17. S. K. Jana, T. Sekac, and D. K. Pal, "Geo-spatial approach with frequency ratio method in landslide susceptibility mapping in the Busu River catchment, Papua New Guinea," *Spat. Inf. Res.*, vol. 27, no. 1, pp. 49–62, 2019, doi: 10.1007/s41324-018-0215-x.
18. A. Zucaro, M. Ripa, S. Mellino, M. Ascione, and S. Ulgiati, "Urban resource use and environmental performance indicators. An application of

decomposition analysis," *Ecol. Indic.*, vol. 47, pp. 16–25, 2014, doi: 10.1016/j.ecolind.2014.04.022.
19. M. Jain, D. Dawa, R. Mehta, and A. P. D. M. K. Pandit, "Monitoring land use change and its drivers in Delhi, India using multi-temporal satellite data," *Model. Earth Syst. Environ.*, vol. 2, no. 1, pp. 1–14, 2016, doi: 10.1007/s40808-016-0075-0.
20. G. Thomson and P. Newman, "Cities and the Anthropocene: Urban governance for the new era of regenerative cities," *Urban Stud.*, vol. 57, no. 7, pp. 1502–1519, 2020, doi: 10.1177/0042098018779769.
21. S. Chen, Y. Zhang, and J. Zheng, "Assessment on global urban photovoltaic carrying capacity and adjustment of photovoltaic spatial planning," *Sustain.*, vol. 13, no. 6, 2021, doi: 10.3390/su13063149.
22. Y. Fan and C. Fang, "Research on the synergy of urban system operation—Based on the perspective of urban metabolism," *Sci. Total Environ.*, vol. 662, pp. 446–454, 2019, doi: 10.1016/j.scitotenv.2019.01.252.
23. M. Li, L. Li, and W. Strielkowski, "The impact of urbanization and industrialization on energy security: A case study of China," *Energies*, vol. 12, no. 11, 2019, doi: 10.3390/en12112194.
24. Y. Guan, L. Kang, C. Shao, P. Wang, and M. Ju, "Measuring county-level heterogeneity of CO_2 emissions attributed to energy consumption: A case study in Ningxia Hui Autonomous Region, China," *J. Clean. Prod.*, vol. 142, pp. 3471–3481, 2017, doi: 10.1016/j.jclepro.2016.10.120.
25. S. Chavanavesskul, "Management of the urban spatial setting to determine the effect of urban heat island on the Bangkok Metropolis, Thailand," vol. 120, pp. 53–63, 2009, doi: 10.2495/SDP090061.
26. S. Garshasbi *et al.*, "On the energy impact of cool roofs in Australia," *Energy Build.*, vol. 278, 2023, doi: 10.1016/j.enbuild.2022.112577.
27. C. Gschwendtner, C. Knoeri, and A. Stephan, "The impact of plug-in behavior on the spatial–temporal flexibility of electric vehicle charging load," *Sustain. Cities Soc.*, vol. 88, 2023, doi: 10.1016/j.scs.2022.104263.
28. A. Bendito, "Grounding urban resilience through transdisciplinary risk mapping," *Urban Transform.*, vol. 2, no. 1, pp. 1–11, 2020, doi: 10.1186/s42854-019-0005-3.
29. X. Xu, H. M. A. Aziz, H. Liu, M. O. Rodgers, and R. Guensler, "A scalable energy modeling framework for electric vehicles in regional transportation networks," *Appl. Energy*, vol. 269, 2020, doi: 10.1016/j.apenergy.2020.115095.
30. L. Sillero, R. Prado, T. Welton, and J. Labidi, "Energy and environmental analysis of flavonoids extraction from bark using alternative solvents," *J. Clean. Prod.*, vol. 308, 2021, doi: 10.1016/j.jclepro.2021.127286.
31. A. Nawab *et al.*, "Exploring urban energy-water nexus embodied in domestic and international trade: A case of Shanghai," *J. Clean. Prod.*, vol. 223, pp. 522–535, 2019, doi: 10.1016/j.jclepro.2019.03.119.
32. M. Mottaghi, H. Aspegren, and K. Jönsson, "Integrated urban design and open storm drainage in our urban environments: Merging drainage

techniques into our city's urban spaces," *Water Pract. Technol.*, vol. 11, no. 1, pp. 118–126, 2016, doi: 10.2166/wpt.2016.016.
33. L. Xiong, S. Li, B. Zou, F. Peng, X. Fang, and Y. Xue, "Long Time-Series Urban Heat Island Monitoring and Driving Factors Analysis Using Remote Sensing and Geodetector," *Front. Environ. Sci.*, vol. 9, 2022, doi: 10.3389/fenvs.2021.828230.
34. S. Jeyasudha, M. Krishnamoorthy, M. Saisandeep, K. Balasubramanian, S. Srinivasan, and S. B. Thaniaknti, "Techno economic performance analysis of hybrid renewable electrification system for remote villages of India," *Int. Trans. Electr. Energy Syst.*, no. April, pp. 1–18, 2020, doi: 10.1002/2050-7038.12515.
35. N. L. Bassuk et al., "On using landscape metrics for landscape similarity search," *Landsc. Urban Plan.*, vol. 117, no. 1, pp. 1–12, 2015, doi: 10.1038/srep11160.
36. A. M. Fathollahi-Fard, L. Woodward, and O. Akhrif, "Sustainable distributed permutation flow-shop scheduling model based on a triple bottom line concept," *J. Ind. Inf. Integr.*, vol. 24, 2021, doi: 10.1016/j.jii.2021.100233.
37. A. Singh and Sushil, "Integrated approach for finding the causal effect of waste management over sustainability in the organization," *Benchmarking*, 2021, doi: 10.1108/BIJ-08-2020-0419.
38. P. Zhang, W. Cai, M. Yao, Z. Wang, L. Yang, and W. Wei, "Urban carbon emissions associated with electricity consumption in Beijing and the driving factors," *Appl. Energy*, vol. 275, 2020, doi: 10.1016/j.apenergy.2020.115425.
39. A. Han, X. Chen, Z. Li, K. Alsubhi, and A. Yunianta, "Advanced learning-based energy policy and management of dispatchable units in smart grids considering uncertainty effects," *Int. J. Electr. Power Energy Syst.*, vol. 132, 2021, doi: 10.1016/j.ijepes.2021.107188.
40. S. F. Afzali, J. S. Cotton, and V. Mahalec, "Urban community energy systems design under uncertainty for specified levels of carbon dioxide emissions," *Appl. Energy*, vol. 259, 2020, doi: 10.1016/j.apenergy.2019.114084.
41. C. Rey-Mahía, L. A. Sañudo-Fontaneda, V. C. Andrés-Valeri, F. P. Álvarez-Rabanal, S. J. Coupe, and J. Roces-García, "Evaluating the thermal performance of wet swales housing ground source heat pump elements through laboratory modelling," *Sustain.*, vol. 11, no. 11, 2019, doi: 10.3390/su11113118.
42. D. Kumar, *Solar Energy Modelling and Assesing Photovoltaic Energy: A Green Energy Initiative*, 1st ed. Gulbarga: Scholar's Press, 2014.
43. M. Žuvela-Aloise, R. Koch, S. Buchholz, and B. Früh, "Modelling the potential of green and blue infrastructure to reduce urban heat load in the city of Vienna," *Clim. Change*, vol. 135, no. 3–4, pp. 425–438, 2016, doi: 10.1007/s10584-016-1596-2.
44. Y. Wu, Y. Shan, S. Zhou, Y. Lai, and J. Xiao, "Estimating anthropogenic heat from an urban rail transit station: A Case study of Qingsheng metro station, Guangzhou, China," *Sustain. Cities Soc.*, vol. 82, 2022, doi: 10.1016/j.scs.2022.103895.

45. M. P. McCarthy, M. J. Best, and R. A. Betts, "Climate change in cities due to global warming and urban effects," *Geophys. Res. Lett.*, vol. 37, no. 9, pp. 1–5, 2010, doi: 10.1029/2010GL042845.
46. S. K. A. Shezan, "Optimization and assessment of an off-grid photovoltaic–diesel–battery hybrid sustainable energy system for remote residential applications," *Environ. Prog. Sustain. Energy*, vol. 38, no. 6, 2019, doi: 10.1002/ep.13340.
47. A. Nutkiewicz, Z. Yang, and R. K. Jain, "Data-driven Urban Energy Simulation (DUE-S): A framework for integrating engineering simulation and machine learning methods in a multi-scale urban energy modeling workflow," *Appl. Energy*, vol. 225, pp. 1176–1189, 2018, doi: 10.1016/j.apenergy.2018.05.023.
48. M. I. Howells, T. Alfstad, D. G. Victor, G. Goldstein, and U. Remme, "A model of household energy services in a low-income rural African village," *Energy Policy*, vol. 33, pp. 1833–1851, 2005, doi: 10.1016/j.enpol.2004.02.019.
49. E. Khorsheed, "Energy load forecasting: Bayesian and exponential smoothing hybrid methodology," *Int. J. Energy Sect. Manag.*, vol. 15, no. 2, pp. 294–308, 2021, doi: 10.1108/IJESM-06-2019-0005.
50. C. Alberto et al., "Overview of Renewable Energy Potential of India," *Model. Earth Syst. Environ.*, vol. 2, no. October, pp. 1–20, 2006, doi: 10.1016/j.compenvurbsys.2015.03.002.
51. International Renewable Energy Agency, "Solar Photovoltaics," 2012.
52. S. J. Quan, J. Park, A. Economou, and S. Lee, "Artificial intelligence-aided design: Smart Design for sustainable city development," *Environ. Plan. B Urban Anal. City Sci.*, vol. 46, no. 8, pp. 1581–1599, 2019, doi: 10.1177/2399808319867946.
53. N. Sihag and K. S. Sangwan, "Development of a sustainability assessment index for machine tools," *Procedia CIRP*, vol. 80, pp. 156–161, 2019, doi: 10.1016/j.procir.2019.01.018.

2
Transitions in the Urban Energy Scenario and Approaches

Deepak Kumar[1,2]

[1]Center of Excellence in Weather & Climate Analytics, Atmospheric Sciences Research Center (ASRC), University at Albany (UAlbany), State University of New York (SUNY), Albany, New York, USA
[2]Amity Institute of Geoinformatics & Remote Sensing (AIGIRS), Amity University Uttar Pradesh (AUUP), Noida, Gautam Buddha Nagar, Uttar Pradesh, India

Abstract

Renewable energy transitions are the subject of research. Renewable energy can be used even in countries that don't have access to the newest renewable energy suppliers thanks to advances in technology. Wind, solar, geothermal, marine, and biomass are all examples of renewable that can be used effectively. Excess power is released and there is a dearth of information on the subject of renewable energy storage. Most international systems run on oil, gas, and coal. Various resources are required to economically and sustainability extract these fuels. In 20 years, energy consumption will increase. Interest in renewable energy sources has increased in response to rising gas prices and worries about the ozone layer. Renewable sources of energy are much less risky than nuclear power. Ozone depletion can be slowed if the harmful impacts of using nonrenewable energy sources are mitigated. Due to rising prices and dwindling oil supplies, energy efficiency measures are becoming more desirable. Potential renewable energy assets are large because they have the potential to outperform energy importance; hence, such resources will have a major proposition in the whole energy portfolio, many of which are focusing on propelling their pool of environmentally friendly power. Emissions from fossil fuels can be decreased thanks to modern technology and renewable energy. There must be a limitless supply of alternatives to oil, and energy efficiency must be improved.

Keywords: Renewable energy, energy storage technologies, solar energy, carbon emission, fossil fuel

Email: deepakdeo2003@gmail.com; ORCID: http://orcid.org/0000-0003-4487-7755

Deepak Kumar (ed.) Urban Energy Systems: Modeling and Simulation for Smart Cities, (19–30) © 2023 Scrivener Publishing LLC

2.1 Introduction

Improving the way power is transferred between the country's many state-run grids will be a crucial step in adapting to the unpredictability of these renewable energy sources. Through increased energy trading between state utilities, it will be possible to achieve a supply of renewable energy that is more consistent and predictable, as well as one that can more effectively supplant expensive and polluting fossil fuels [1, 2]. This will be possible because periods of surplus electricity in one region can compensate for deficiencies in electricity supply in other regions. The low prices of wind and solar installation, on the other hand, disguise the structural costs that renewable energy creates for grids when it is implemented at larger proportions [3–5]. This is because wind and solar installations are typically located in remote areas where there is limited access to electricity. The intermittent and unpredictability of the electricity generated by such technologies, in contrast to the constant and on-demand flow of power from traditional fossil fuel generation, necessitates fundamental changes in the way in which nations invest in and operate their grids. These changes are necessary because the intermittent and unpredictability of the electricity generated by such technologies is a significant challenge. These adjustments are necessary to accommodate the nature of the electricity provided by such technologies and are therefore required [6–9].

Therefore, India will need to make more substantial adjustments to its grid management to accommodate fluctuating renewable energy sources like wind and solar. Integration of the country's many state-run electrical grids into a more coordinated system that may use surplus electricity in one region to compensate for inadequacies in another will be an essential but difficult solution. This allows renewable energy sources that fluctuate in output to replace more expensive and polluting fossil fuels more reliably while still providing enough service to a wider area [10–12].

Maintaining India's rapid expansion of renewable energy through deeper reforms would provide clean, affordable power to its people and help the world offset the failings of more laggard countries, bringing the Paris Agreement's more aggressive target of keeping global warming to 1.5 degrees Celsius back within reach. But overcoming the recurring cycles that have plagued India's power sector for decades is still to come [13, 14]. However, because India's current method for controlling its grid is not the most effective one, it is possible that this synergy will not be able to sustain itself. The bigger the quantity of wind and solar power that the country installs, the more difficult and expensive it will be to use such energy. This poses a risk to the transition that the country is making to renewable energy since it threatens to put the brakes on the transition [15–17]. The intermittent nature of renewable energy sources such as solar and wind power presents a challenge that is not encountered with conventional kinds of generating with fossil fuels. Solar panels can only generate electricity when the sun is shining, and wind turbines can only generate electricity when the wind is blowing, but the generation of electricity using coal and natural gas can produce a consistent and ongoing supply of electricity. As a consequence of this, it is typical for wind and solar power facilities to go dark for extended stretches during the day and throughout the year. Although the majority of this volatility does follow recognized patterns—for example, the output of solar power is at its highest when the sun is at its zenith, and the output of wind power in India is at its highest during the monsoon season—its precise size is impossible to predict [1, 18]. The fact that state utilities have even stronger incentives to prioritize shorter-term interests is the most fundamental barrier to regional coordination. This can take the form of state utilities choosing to maintain individualized control over energy dispatch rather than committing to a system that is more independent and centralized, regardless of how much money saving through regional coordination might save in the long run [19, 20]. Because of decades of financial mismanagement and bribery, this chasm exists because India's electricity businesses have been utilized more as political pawns than reliable utilities. This has been the case for many years. In the early years of India's existence as a nation, one of the state's primary objectives was to make certain that its population was healthy [21, 22]. The trust of some individuals is a problem. Customers need to have faith that governments will deliver on their promises of greater quality power in the not-too-distant future because investing in electricity infrastructure frequently necessitates immediate price rises but may take years to show any results. This is because the investments can take years to show any results. On the other hand, governments typically fail to plan on such a long-term horizon either because they lack the requisite self-control to do so or because they are simply frightened that people will not continue to be committed to such an arrangement. A considerable number of would-be reformers have been removed from office before their work could be completed, only to watch the subsequent ruling party go back to the traditional system of cross-subsidies once they have been replaced. Because of this, the state of the power sectors in the majority of states has, for the most part, stayed much the same over several decades.

2.2 Recent Transformation in Energy Sectors

These climate efforts have made India a leader in the sustainable energy transition. This may be observed in the volume and velocity of wind and solar installation, the quick acceptance of electric vehicles, and the rapid expansion of renewables and battery manufacturing. While these energy transformations are critical pillars of India's attempts to address climate change, the broader economic shift is also important. India's climate and renewable energy measures are based on growing domestic industry, creating local jobs, securing supply chains, and bolstering energy security [23, 24]. For India's decarbonization journey to align with inclusive economic growth, the country must accelerate infrastructure reform and navigate federal politics, land, and labour systems, as well as secure low-cost transition finance. The time has come for developed economies to step up and support these efforts, with current estimates estimating India will require more than $17 trillion in investments to reach its net-zero targets. Despite this setback, India is pressing ahead with its innovative course, from which other growing countries can learn and draw inspiration.

It's remarkable to watch how quickly India is evolving in this day and age. Its economic growth over the past two decades has been among the highest in the world, which has made it possible for millions of people to climb out of poverty. The rate of urbanization in India is equivalent to adding a city the size of London every year, which calls for an enormous amount of investment in the construction of new infrastructure including roads, factories, and transportation systems [25, 26]. Coal and oil have been the backbones of India's economic progress and modernization for the most part. As a result, more and more individuals in the country can obtain access to reliable and convenient modern energy sources. This has included providing new electrical services to an average of 50 million people annually over the past decade. India confronts numerous immediate hurdles despite its ambitious goal of reaching net zero. With the steep rise in commodity prices, energy has become more expensive, and the tightening of markets has raised concerns about energy security for the world's third-largest energy importer. Many people still do not have access to a consistent electrical source. Many people's health is being negatively affected since they still use traditional fuels for cooking. The urgent restructuring of the electricity distribution industry is being hampered by financially struggling electrical distribution businesses. And because of the high levels of pollution, the air quality in Indian cities is among the worst in the world. India's plans to combat and adapt to climate change are revolutionary not just for the country, but for the entire world. Together, NITI Aayog and the IEA will help India grow, industrialize, and improve the lives of its population without increasing carbon emissions. There are a lot of solid reasons for onshore wind to play a larger role in India's electrical infrastructure [27, 28]. To provide a clean and affordable alternative to fossil fuels, projects are being constructed at a cost that is lower than that of a new coal plant, the country's principal source of power. The ration card system is meant to help the poor by making kerosene more affordable; however, it is not very well targeted. The wealthy and middle class stand to gain the most from this. Without an address, the poor cannot get a ration card and must pay significantly more on the black market for kerosene or use inefficient and expensive fuelwood for cooking. However, kerosene is only allocated to ration stores intermittently and in small quantities, so it frequently runs out before the next allotment, forcing the poor to either buy kerosene on the open market or utilize fuel wood.

Poor power quality slows economic growth, and the power sector's disastrous finances drain public cash that could be utilized for targeted investments or welfare. The dysfunction of India's discoms is one of the biggest impediments to modern economic progress. As long as India's distribution businesses (discoms) are financially insolvent, politically compromised, and sheltered from the incentive to earn a profit, they will hinder the country's capacity to reach its renewable energy ambitions, starting with regional cooperation. Almost half of all residential energy costs go toward electricity, and the proportion is even higher in homes with higher incomes. Despite a general decrease in household income, there has been an increase in the prevalence of electric appliances since the last surveys were conducted. This may be the result of a combination of factors, including falling prices for some of the appliances in question and a shift in priorities that places greater importance on leisure and comfort spending. Electricity is becoming increasingly important in the home, yet it is also the most unreliable of the primary energy sources. In the dry season, when it is essential, electricity is often off for a few hours a day due to frequent brownouts. To ensure adequate supply in the future, a variety of policies, including but not limited to, increased pricing, privatization of distribution, and the use of more energy-efficient appliances and lighting will be required, but so far, the political and organizational will to enforce such measures has been lacking.

2.3 Research Progressions

Urban energy modelling has many issues. These issues include model complexity, data uncertainty, model integration, and policy relevance. New data sources including GPS/GSM signal traces, sensitivity analysis, cloud computing, and activity-based modelling can help reduce some of these challenges. By capturing more policy levers, resolving uncertainty, and setting standards for consistent and comparable data, we intend to improve the policy relevance of this field's work. To safeguard India's astonishing transition away from fossil fuels and toward more environmentally benign renewable energy, the usage of renewable energy sources has expanded at a dizzying rate in recent years, even though the government has not provided adequate aid for the business. India has doubled its wind and solar power in five years. This surge was motivated by clean energy's promise to address people's critical development requirements efficiently and cost-effectively.

Urban energy systems modelling will likely combine activity and agent-based methodologies, improved data standards, and computational advances to estimate geographically and temporally disaggregated resource demands within an integrated model of urban energy supply. Such a model could represent the complexity of modern cities and allow the examination of policy packages that span once separate but now interdependent domains like transportation, power generation, heating and cooling, and water delivery. Without a doubt, there is a lot of potential in modelling urban energy systems, and there are also many other obstacles, such as the difficulties of modelling growing economies, that aren't discussed here. However, the purpose of this work has been to attempt to synthesize a wide variety of approaches into a coherent whole, one to which academics and industry professionals alike can contribute and advance the state of the art.

The development of both industry and technology has contributed to a rise in the demand for an increase in available power. A broad prediction of the need for alternative and augmented energy resources has been motivated by the scarcity and depletion of other non-renewable resources [29–31]. In recent years, renewable energy sources like wind and solar have emerged as potential clean energy options. Thus, the current focus has shifted to increasing the size of energy-generating hybrid systems. There have been many attempts to show how technical progress meets regional needs. Though studies have begun to assess the prototype's performance, they have received almost any focus. An additional essential component of this work is the simulation of hybrid urban renewable energy systems to research the technological and economic feasibility of the system. Previous attempts were made to report the technological, scientific, and industrial progress that has occurred because of the hybrid renewable energy system. An hourly load profile of a typical urban requirement was modelled to assess the metropolitan zone's energy potential [29, 32, 33]. These will detail past, current, and future developments in the design, development, and implementation of hybrid energy systems for the metropolitan region, which can subsequently be exported.

There is a need to create carbon markets in India to encourage the use of renewable energy sources and technological advancements. Most companies today use open access strategies, which entails them purchasing green energy directly from suppliers. When implemented on a large scale, such methods might add complexity to the grid. Consequently, measures that incentivize decarbonization and satisfy the complex needs of the energy system through market-based prices are required, in addition to legislative initiatives like green open access.

It is necessary to quicken the pace of growth in the energy sector to support the overall economic boom in India, despite the country's efforts to decarbonize its economy by boosting renewable energy, electrifying transportation systems, and assuring energy efficiency. To achieve its environmental goals and economic objectives, the government must construct a robust clean energy infrastructure built on the pillars of reliability, sustainability, independence, and affordability.

2.4 Breaking the Cycle

Understanding, administering, and building cities is the goal of an interdisciplinary field like urban information processing, which draws on recent advances in computing and communication to inform systematic ideas and approaches. Multiple aspects of urban infrastructure are covered by this field. Transportation, housing, retail, physical infrastructure involving waste, water, electricity, and other types of energy, demographic and economic geography, urban development, and many other aspects of cities and urban systems are studied independently. Legal residents can use ration cards to buy subsidized kerosene. Because the government doesn't distribute ration cards to unlawful squatter communities, low-income consumers must buy kerosene at higher costs or burn fuelwood or charcoal. Each family was once limited to one subsidized LPG cylinder, which may be switched when empty. Although India's efforts to tackle climate change have centred on growth rather than sustainability, this may alter as climate change's repercussions become more obvious. Climate change is sometimes referred to as a "public good," indicating that any action taken to reduce emissions benefits all nations. As a direct consequence of this, we now have access to streams of data on the operation of a city in real time that were not available to us before. These streams of data were previously collected manually rather than automatically through sensors. As a result, there is now a phenomenon known as "big data," which consists of records created in real time that are both extremely diverse and extremely numerous. Sensors that run nonstop and report live updates to the system in question could be the source of such data. We need novel approaches and models for this data to better comprehend them and make sense of the old models that may be usefully interpreted considering the new information. Now that temporal dynamics is a serious part of this discipline of informatics, this has pushed the topic of the 24-hour city to the forefront. We used to put more emphasis on spatial diversity in our models, but now we're also taking time into account. Clean energy technologies have the potential to accelerate the development of low-cost carbon energy infrastructure, which in turn is accelerating the development of renewable energy technologies. The current research considered the various limitations before giving its stamp of approval for the concept's implementation. It was assumed that providing electricity to people in undeveloped countries is a complex task with few viable options for a solution. Energy subsidies have wide-ranging effects that extend beyond the budgets of individual governments. These effects can be seen in the financial markets, in society, and the natural world. Energy subsidies are also connected to issues such as improving access to energy, public health, and climate change. The public and those responsible for formulating policy should be able to engage in an educated discussion on the degree to which their actions are in line with India's goals if there is transparency. Energy systems modelling simulations show that India can safely contemplate increasing the percentage of renewables used in electricity generation as part of its strategy to achieve a future with zero net emissions. A zero-carbon electricity system, according to energy system modellers, must be financially sustainable and supported by strong storage technology. Models of the energy system are now being worked on to become more transparent, considering local objectives and realities, and incorporating equity. Making progress toward India's zero-net-energy target requires significant investment and innovative technology. Multiple approaches may help the government attain its net-zero objective, depending on technological factors. For commercial use, cost-cutting innovations are needed. Politicians should provide incentives for innovation rather than try to predict the future climate or regulate technology and power sources.

2.5 Conclusion

Cities in underdeveloped countries are gradually switching from using biomass-based fuels to using modern fossil fuels and electricity. The second group of energies is easier to integrate into daily life and their use results in fewer environmental impacts. The fuel shift has occurred in tandem with the expansion of urban economies. Researchers are collecting empirical evidence to back up these broad claims. The survey's findings about the underlying fuel shift are consistent with those found in other urban centres of emerging countries. However, the government's policies have favoured the wealthy due to disparities in access to resources like gasoline and machinery. We present the survey's key findings and place them alongside other statistics for India, then propose policy measures to address the resulting inequalities. Energy models are crucial for the practical implementation of energy transitions because they are mathematical instruments that depict different energy-related problems as accurately as possible. A recent trend analysis of energy system models discovered an increased emphasis on open access, modelling cross-sectoral synergies, and greater temporal detail. This was done to deal with the challenge of planning for future scenarios that involve a high amount of variable renewable energy (RE) sources. However, "major issues remain in expressing high-resolution energy demand across all sectors," as well as in "openness and accessibility," "how tools are coupled together," and "the level of collaboration between tool developers and policy/decision-makers."

2.6 Future Implications

Until the price of battery storage technologies drops to where they can compete with traditional fossil fuel power plants, a small amount of non-RE-based electricity generation will be necessary. However, according to the above-mentioned trend analysis, this will be possible in the not-too-distant future due to the rapidly falling prices of lithium-ion batteries. When it comes to energy-related greenhouse gas emissions, the power sector in India has consistently been the major contributor. According to the research, "air pollution, predominate in cities, has stemmed from the reliance on low-quality coal used in very inefficient power plants, exacerbating other environmental difficulties." Several significant policy measures, such as the green hydrogen policy, the offshore wind policy, the promotion of electric vehicles, and the establishment of a green day-ahead market, reflect the government's dedication. Coal and crude oil remain India's key energy sources. Renewable energy would cost a lot to replace. The administration has proved its commitment through numerous significant policy actions, including green hydrogen, offshore wind, electric vehicle promotion, the introduction of a green day-ahead market, and the relaxing of open access to green energy. Coal and crude oil remain India's key energy sources. Replace them with green energy.

References

1. A. Das and V. Balakrishnan, "Grid-connectivity of remote isolated islands-A proposition in Indian context," *J. Renew. Sustain. Energy*, vol. 4, no. 4, 2012, doi: 10.1063/1.4738591.
2. A. Sarin, R. Gupta, and V. V. Jituri, "Solar Residential Rooftop Systems (SRRS) in South Delhi: A strategic study with focus on potential consumers' Awareness," *Int. J. Renew. Energy Res.*, vol. 8, no. 2, pp. 954–963, 2018.
3. M. A. Rahman, I. Rahman, and N. Mohammad, "Demand side residential load management system for minimizing energy consumption cost and reducing peak demand in smart grid," in *2020 2nd International Conference on Advanced Information and Communication Technology, ICAICT 2020*, 2020, pp. 376–381, doi: 10.1109/ICAICT51780.2020.9333451.
4. Y. Shi, D. Feng, S. Yu, C. Fang, H. Li, and Y. Zhou, "The projection of electric vehicle population growth considering scrappage and technology competition: A case study in Shanghai," *J. Clean. Prod.*, vol. 365, 2022, doi: 10.1016/j.jclepro.2022.132673.
5. A. Bugaje, M. Ehrenwirth, C. Trinkl, and W. Zörner, "Electric two-wheeler vehicle integration into rural off-grid photovoltaic system in kenya," *Energies*, vol. 14, no. 23, 2021, doi: 10.3390/en14237956.
6. M. I. Howells, T. Alfstad, D. G. Victor, G. Goldstein, and U. Remme, "A model of household energy services in a low-income rural African village," *Energy Policy*, vol. 33, pp. 1833–1851, 2005, doi: 10.1016/j.enpol.2004.02.019.
7. A. Chow, S. Li, N. Poursaeid, A. Fung, and C. Xiong, "Urban Solar Energy Modeling & Demonstration Technology," in *EIC Climate Change Technology Conference 2013*, 2013, no. 1569694731, pp. 1–13.
8. V. G. Dovì, F. Friedler, D. Huisingh, and J. J. Klemeš, "Cleaner energy for sustainable future," *J. Clean. Prod.*, vol. 17, no. 10, pp. 889–895, 2009, doi: 10.1016/j.jclepro.2009.02.001.
9. E. Khorsheed, "Energy load forecasting: Bayesian and exponential smoothing hybrid methodology," *Int. J. Energy Sect. Manag.*, vol. 15, no. 2, pp. 294–308, 2021, doi: 10.1108/IJESM-06-2019-0005.
10. H. K. Rn, G. M. Shafiullah, A. Maung, T. Oo, and A. Stojcevski, "Feasibility study of Standalone Hybrid Power system modeled with Photovoltaic

modules and Ethanol generator for Victoria," *Int. J. Eng. Sci. Invent.*, vol. 3, no. 8, pp. 54–69, 2014.
11. D. P. Boon et al., "Groundwater heat pump feasibility in shallow urban aquifers: Experience from Cardiff, UK," *Sci. Total Environ.*, vol. 697, 2019, doi: 10.1016/j.scitotenv.2019.133847.
12. Rena, S. Yadav, S. Patel, D. J. Killedar, S. Kumar, and R. Kumar, "Eco-innovations and sustainability in solid waste management: An indian upfront in technological, organizational, start-ups and financial framework," *J. Environ. Manage.*, vol. 302, 2022, doi: 10.1016/j.jenvman.2021.113953.
13. G. von Högersthal, A. Lui, H. Tomičić, and L. Vidovic, "Carbon pricing paths to a greener future, and potential roadblocks to public companies' creditworthiness," *J. Energy Mark.*, vol. 13, no. 2, pp. 1–24, 2020, doi: 10.21314/JEM.2020.205.
14. I. Marques-Perez, I. Guaita-Pradas, A. Gallego, and B. Segura, "Territorial planning for photovoltaic power plants using an outranking approach and GIS," *J. Clean. Prod.*, vol. 257, 2020, doi: 10.1016/j.jclepro.2020.120602.
15. S. Mukherjee and A. Asthana, "Techno-Economic Feasibility of a Hybrid Power Generation System for Developing Economies," *Proceedings*, vol. 1, no. 7, p. 693, 2017, doi: 10.3390/proceedings1070693.
16. A. Iqbal and T. Iqbal, "Design and Analysis of a Stand-Alone PV System for a Rural House in Pakistan," *Int. J. Photoenergy*, vol. 2019, p. 8, 2019.
17. V. Laroche, P. Pasquier, and B. Courcelles, "Integration of standing column wells in urban context: A numerical investigation-case in the City of Montreal," *Sustain. Cities Soc.*, vol. 78, 2022, doi: 10.1016/j.scs.2021.103513.
18. A. Sharma, "Potential of Non-conventional Source of Energy and Associated Environmental Issue: An Indian Scenario," *Int. J. Comput. Appl. (0975 – 8887) Recent Trends Electron. Commun. (RTEC 2013)*, no. Rtec 2013, pp. 43–47, 2017.
19. Z. Zheng, Z. Wu, Y. Chen, Z. Yang, and F. Marinello, "Analyzing the ecological environment and urbanization characteristics of the Yangtze River Delta Urban Agglomeration based on Google Earth Engine," *Shengtai Xuebao/ Acta Ecol. Sin.*, vol. 41, no. 2, pp. 717–729, 2021, doi: 10.5846/stxb202003250687.
20. A. M. Ellison, "Macroecology of mangroves: Large-scale patterns and processes in tropical coastal forests," 2002, doi: 10.1007/s00468-001-0133-7.
21. S. Völker and T. Kistemann, "Developing the urban blue: Comparative health responses to blue and green urban open spaces in Germany," *Heal. Place*, vol. 35, pp. 196–205, 2015, doi: 10.1016/j.healthplace.2014.10.015.
22. M. Ariken, F. Zhang, K. Liu, C. Fang, and H. Te Kung, "Coupling coordination analysis of urbanization and eco-environment in Yanqi Basin based on multi-source remote sensing data," *Ecol. Indic.*, vol. 114, no. January, p. 106331, 2020, doi: 10.1016/j.ecolind.2020.106331.
23. M. Mukherjee and K. Takara, "Urban green space as a countermeasure to increasing urban risk and the UGS-3CC resilience framework," *Int. J. Disaster Risk Reduct.*, vol. 28, pp. 854–861, 2018, doi: 10.1016/j.ijdrr.2018.01.027.

24. P. Choudhary and R. K. Srivastava, "Techno-economic case study: Biofixation of industrial emissions at an Indian oil and gas plant," *J. Clean. Prod.*, vol. 266, 2020, doi: 10.1016/j.jclepro.2020.121820.
25. S. Y. Irda Sari, D. K. Sunjaya, H. Shimizu-Furusawa, C. Watanabe, and A. S. Raksanagara, "Water sources quality in urban slum settlement along the contaminated river basin in Indonesia: Application of quantitative microbial risk assessment," *J. Environ. Public Health*, vol. 2018, 2018, doi: 10.1155/2018/3806537.
26. W. Xiaohua and F. Zhenmin, "Rural household energy consumption with the economic development in China: Stages and characteristic indices," *Energy Policy*, vol. 29, pp. 1391–1397, 2001, doi: 10.1016/S0301-4215(01)00037-4.
27. F. Rajabiyazdi, G. A. Jamieson, and D. Q. Guanolusia, "An Empirical Study on Automation Transparency (i.e., seeing-into) of an Automated Decision Aid System for Condition-Based Maintenance," *Lect. Notes Networks Syst.*, vol. 223 LNNS, pp. 675–682, 2022, doi: 10.1007/978-3-030-74614-8_84.
28. C. Ruiz and M. Quaresma, "Explainable AI for Entertainment: Issues on Video on Demand Platforms," *Lect. Notes Networks Syst.*, vol. 223 LNNS, pp. 699–707, 2022, doi: 10.1007/978-3-030-74614-8_87.
29. K. Dheeraja, R. Padma Priya, and T. Ritika, *Optimal Real-time Pricing and Sustainable Load Scheduling Model for Smart Homes Using Stackelberg Game Theory*, vol. 99, 2022.
30. D. Kumar, "Urban energy system management for enhanced energy potential for upcoming smart cities," *Energy Explor. Exploit.*, p. 014459872093752, Jul. 2020, doi: 10.1177/0144598720937529.
31. A. Gupta, S. Mishra, N. Bokde, and K. Kulat, "Need of smart water systems in India," *Int. J. Appl. Eng. Res.*, vol. 11, no. 4, pp. 2216–2223, 2016.
32. N. Thirumurthy, L. Harrington, D. Martin, and L. Thomas, "Opportunities and Challenges for Solar Minigrid Development in Rural India Opportunities and Challenges for Solar Minigrid Development in Rural India," no. September, 2012.
33. T. Kugler *et al.*, "Sustainable Heating and Cooling Management of Urban Quarters," *Sustain.*, vol. 14, no. 7, 2022, doi: 10.3390/su14074353.

3

Urban Renewable Energy Resource Optimization Systems

Kalpit Jain[1]* and Devendra Kumar Somwanshi[2]†

[1]Department of Mechanical Engineering, Sangam University, Bhilwara, Rajasthan, India
[2]Department of Electronics and Communication Engineering, Poornima College of Engineering, Jaipur, Rajasthan, India

Abstract

The need for a renewable energy source that won't harm the environment is growing. By 2050, energy demand is projected to rise considerably. Most energy needs are met by petroleum derivatives, although green electricity can aid. Renewable energy comes from the sun, the ground, or the environment. Renewable energy is "electricity and heat created from solar, wind, sea, hydropower, biomass, geothermal assets, bio energizes, and hydrogen." The creation of renewable energy comes from various sources like long-term sustainable Solar PV generation, BIPV optimization, solar resources, wind energy, hybrid solar-wind power generation, solar thermoelectric hybrid systems and microgrid systems. Microgrid systems are arising as a suitable option as being less reliant upon brought together energy supply and can here and there utilize more than one fuel source. Decentralized neighborhood frameworks including those utilizing nearby assets of renewable energy like wind, solar and biomass, appear much more feasible. The degree of microgrid additionally decides the condition for the framework to be worked in either grid-connected (GC) or stand-alone (SA) mode. Renewable energy (RE) is normal as an ideal answer for decreasing a worldwide temperature alteration and embracing a practical turn of events. Various kinds of techniques are utilized in various circumstances, for example, biogas and biomass innovation, sunlight-based photovoltaic plants, wind turbines, micro-hydro technology, and hybrid systems by different scientists. Power supplied by enhancing locally available renewable sources of energy is likely to benefit a wide variety of areas in India.

---

*Corresponding author: er.kalpit1988@gmail.com; https://orcid.org/0000-0001-9102-4566
†Corresponding author: imdev.som@gmail.com; https://orcid.org/0000-0003-4331-0917

Deepak Kumar (ed.) Urban Energy Systems: Modeling and Simulation for Smart Cities, (31–46) © 2023 Scrivener Publishing LLC

To plan and carry out mediation efforts like this, concrete evidence of niches that would directly benefit from such a system is required, as is the exploration and evaluation of all viable options for power supply. In addition to conducting a comprehensive audit of India's energy supply situation, it is crucial to direct energy studies in some un-energized towns and assess the availability of renewable energy assets in the same. Appropriate renewable energy-based systems will be established, and their yearly power conveyance evaluated based on the findings of energy studies and other supplementary information collection. The proposed research has the potential to recommend feasible designs for hybrid frameworks that incorporate diverse energy resources for use in unserved communities, and the technique developed for this purpose can be applied to similar situations in other parts of the country.

Keywords: Renewable energy, wind energy, microgrid, biogas, sunlight

3.1 Introduction

A vital part of our cutting-edge human advancement is the energy of various structures. Without energy, several key functions may be rendered inoperable. So it may be said that energy is a component of our lives. There are many distinct kinds of energy utilization cycles, the most notable of which include the exploitation of the raw energy that can be found in falling water, in stocks of coal, oil, gas, and so on [1].

The energy that comes from renewable sources is generated through recurrent cycles that are continuously updated. In its various forms, it obtains its energy either directly from the sun or indirectly from the heat that is generated deep inside the ground [2]. Power and warmth are produced from sunlight-based, wind, sea, hydropower, biomass, geothermal assets, bio fills and hydrogen got from renewable resources. It is important to keep in mind that power and warmth obtained from solar, wind, marine, hydroelectric, biomass, geothermal, and biofuel resources, as well as hydrogen, are considered to be examples of renewable resources [3].

Renewable sources of energy include hydropower, solar energy, wind energy, flowing energy, biomass fill, and so on. These are all examples of fuel sources that can be retrieved after a typical time cycle has passed. There are certain regions where access to renewable energy sources is restricted, and the extent of that restriction is contingent on the traits that distinguish each source [4]. When renewable energy sources are extracted at a pace that is higher than their rate of regeneration, the energy no longer qualifies as renewable. The power comes from resources and technologies that don't deplete the planet's supply of clean energy.

The current energy demand is gradually increasing for various reasons, including a growing population, a desire for more productivity, an expectation of higher levels of day-to-day comfort, and overall advancements in financial and mechanical technology [5]. The power age framework relies upon imported oil and its gaseous petrol. Then again, as the data about the stores of petroleum products, if they are devoured at the current rate, the turned-around flammable gas and coal will be depleted constantly from 2020 to 2030.

To reduce the dependency on imported fuel and the pressure on natural gas, the present power generation system must be diversified and at the same time, indigenous energy resources must be explored and developed. It may be mentioned that concern for the environment is now a universal issue and conventional energy gives rise to greenhouse gases with adverse consequences for health and climate [6]. From these points of view, the harnessing of renewable energy sources and the advancement of relative innovations is an exceptionally significant choice. Networks in rural areas and principally in distant regions have next to no conceivable outcomes to take part in the public power supply.

Hardly any researchers examined the capability of wind energy. Most analysts worked for expanding the energy created by wind. Reconciliation of wind energy into the existing power framework causes specialized difficulties like voltage guidelines, steadiness, and force quality. Various methodologies like battery storage and diesel generator sets were recommended for expanding the security and working on the productivity of wind energy [7].

Many scientists have not examined the crossbreed Solar-Wind Energy Generation and did recreations on HOMER and the outcomes got exhibited as energized results which were superior to individual sun-oriented or wind energy frameworks.

Specialists didn't suggest the utilization of explicit techniques which effectively affected person/soil/harvests and climate. No specific work was found on Efficient Methodology for Biogas age [8].

3.2 Literature Review

3.2.1 Long-Term Sustainable Solar Power Generation

[**Alfred H. Canada, 1995**] proposed the plan and application for the regular dispatch competent force from a 100MW SVC power plant and connected to the utility plant was presented together with the plan and application for a 100MW to 10KW solar facility for the climatic conditions of the Pacific Northwest. The authors completed designs, implementation, and operation of low-cost SVG power plants and established interface with any autonomous force manufacturers PV module for electrical utilities [1].

[**Rsuzuki, et al., 2002**] assumed the illustrative approach for a PV framework to pick the unlucky elements impacting the efficacy of becoming older: The authors isolated a direct and diffuse portion of sun-based irradiance to investigate its effects, as well as the location of occurrence and the east/west azimuth. In order to evaluate the temporal variation in power output, the authors hypothesized that they could separate the effect of direct and indirect solar irradiance at a given episode point [2].

[**Zhenhua Jiang, 2006**] proposed a plan to provide constant high-quality electricity; the executives were presented with a plan for a hybrid system comprised of energy units and batteries coordinated with PV power frameworks. The author emphasized that a potential drawback of energy components is the potential for their low elements to reduce yield voltage during load pick. Increasing the size of these energy components above their norm is necessary to achieve maximum power output and thereby victory [3].

[**Hassan Moghbelli, et al., 2006**] suggested implementing mobile energy in the photovoltaic converter and weighed the influence on growing the yield power attributes against the fixed energy's impact. The sun's constant motion meant that the uneven distribution of sunlight that reached the ground was incompatible with a reliable source of PV power. As the sun moves across the sky, the suggested flexible PV energy [4].

[**Lander H. P., et al., 2009**] addressed the unique constraints of renewable energy development on UPERM and the progress that has been made so far. The authors emphasized the detrimental practical and natural impact of the energy age's dependency on petroleum goods, which was highlighted. Potential energy age assets were discussed, along with the factors that may prevent their realization. These factors include, among others, the high cost of PV modules and the high quality of PV cells [5].

[**DAI Qinghui, et al., 2009**] led the research and study of the methods presented to improve the effectiveness of solar-based force. It is challenging to progress solar energy for a broad scope because of the high cost of PV boards and the low flow productivity of converting solar energy into electricity [6].

[H. J. Hou, et al., 2009] showed Analysis of Concentrating Solar Power for Power Generation from Renewable Sources. To generate reliable power from solar energy with little financial and environmental impact, the use of concentrators is rapidly becoming the technology of choice. Heliostats, the recipient, the steam age framework, and the capacity framework are all crucial parts of the solar-powered apex. Solar-powered apexes have substantial potential in the long run for exceptionally high efficiency rates of transformation. Dish-shaped reflectors, or "explanatory dish structures," are typically quite compact. Every new development has both useful applications and undesirable drawbacks. While the climate is ideal for constructing CSP facilities, significant barriers to implementation include the high capital expense and innovative nature of PSP systems. However, in the long run, CSP will get competitive with conventional power plants due to the leveled-out energy cost decline and advanced innovation [7].

[K. C. Mcintyre, et al., 2010] replicating PV panels and energy units to create a small network is being explored. Optimizing Integrated Models through Simulation. It was also considered that basic amenities like running water, sewage, natural gas, and solar panels/wind turbines would be readily available. From the HOMER simulation results, it was clear that connecting the PV cells to the existing network was the optimal setup. Considering the high capital speculation and substitution cost, simulation findings indicated that improvements in energy components and Microturbine technology should have been made [8].

[Xiaoli Xu, et al., 2010] evaluated the internet's informational and quality bounds of the engine architecture across all climates using a climate-responsive auto-following approach. Creative PV cells were used in solar power-based force generation, but their performance was inconsistent. The shortcomings of previous systems, such as inaccurate sun-oriented azimuth following, inefficient force aging, limited adaptability, and high cost, are addressed and resolved by these new designs. The framework centralizes solar-based azimuth light sensation following and time following auto-following climate data in one place. The findings of the experiments demonstrate that the use of a climate-independent auto-following technique improves the precision and dependability of solar-powered activities. The power age framework improves photograph electric change efficiency and hostility to twist capacity of the solar-based cell plate while decreasing the power dispersal of the force age system [9].

[Yahyavi, *et al.*, 2010] studied the effect of height on framework output and determined whether force and energy per unit volume could be reliably estimated. The authors investigated how variations in solar radiation affect the performance of a solar energy power generation system. At the time, it was believed that using solar energy to maintain a healthy environment was a viable choice. The authors conclude that using trackers increases energy output by 57%, and that it is more efficient to equip the system with a tracker and raise the construction of sunlight-based force for a wider range of sun energy per unit volume [10].

[Yusuke, *et al.*, 2010] did an investigation to diminish the complete expense of energy and decrease CO_2 discharge by introducing PV and BESS (battery energy stockpiling system) into request side like a production line, business buildings and retail plazas. In reproduction, the forecasting of burden and PV age power are incorporated by the expense and CO_2 outflow may change by changing the blunders of determining. The enhancements of cost and CO_2 outflow change by changing the exactness of estimating the decrease by introducing BESS are gotten if the precision of anticipating was high. The decrease in the cost was gotten by the heap evening out if the precision of forecasting was high. Aftereffects of reproduction show a 10-15% decrease in cost on PV and 2-3% in BESS and CO_2 [11].

[**Huang Chao-Hui**, *et al.*, 2011] conveyed an analysis utilizing miniature lattice programmed age control utilizing fluffy framework and neural framework. The miniature lattice contains fans, sun-based cells, diesel motors and a battery and simulation was done in MATLAB. MATLAB results showed that this programmed age control framework can handle each miniature force, diminish the utilization of diesel motors and expand wind and sun-based energy [12].

[**Aakanksha Aggarwal**, *et al.*, 2011] proposed an instrument to accomplish the most extreme sun-oriented energy by painting the steel sheets with sunlight-based paint which was fixed on the dividers of the structure with an adaptable boating point. These would resolve the significant issue of draining fuel sources [13].

[**Sharaf**, *et al.*, 2012] proposed a novel hybrid calculation to conjecture momentary PV power yield. The photovoltaic force was a substitute for regular energy and was ecologically well disposed; however, because of the wild volume of PV power yield was a significant test while associating it to the network. The author proposed a calculation consolidating frequency change (WT), fluffy ART MAT (FA) organization and firefly (FF) to gauge PV yield [14].

[**Bonifride Tuyishimire, *et al.*, 2013**] fostered a different rate Kalman indicator that gives continuous anticipating of sun-based PV age. In sun-based PV the electrical age and transmission dissemination depended on unidirectional changeability and non-dispatch capacity attributes of sun-based energy, making it trying to keep up with significant degrees of framework dependability. Kalman filters are utilized for recurrence space demonstrating strategy and for giving beginning preparing qualities to ANN [15].

[**Esko Juuso, *et al.*, 2013**] clarified the effect/impact of Linguistic Equation (LE) regulators in sun-oriented energy. Because of cyclic varieties and barometrical conditions like overcast climate mugginess and air straightforwardness, the effectiveness of sunlight-based energy age was questionable. A quick beginning up and effective activity in changing overcast conditions were vital. Different techniques like a feed-forward regulator, PID regulator, interior model regulator and GA had been utilized for multi-target tuning LE utilized in numerous businesses for calibrating LE regulators utilize model-based connectors and feed-forward highlights which forestall overheating and uncommon highlights [16].

[**Vaijukal khambkar *et al.*, 2014**] proposed discretionary measuring of sunlight-based PV and battery blend in a lattice-associated framework. An insightful strategy for measuring sun-based PV and battery was proposed to lessen the line misfortunes and irregularity of sun-powered PV and battery to handle the fluctuation which was anticipated by likelihood thickness work and upheld battery energy stockpiling. The area and estimating approach was applied utilizing MATLAB. With the proposed ideal estimating procedure the misfortunes can be decreased by up to 83% during which the sun-based PV and battery were associated [17].

[**Jinn Chang Wu, *et al.*, 2014**] Another solar-powered force age architecture using a dc-dc power converter and a seven-level inverter was proposed by the same group of authors in 2014. They used a capacitor selection circuit and a complete scaffold power connector to create the new seven-stage inverter. The dc-dc power converter's circuit converts the two yield voltage wellsprings' three-level dc voltage and the full scaffold converter's three-level dc voltages into the seven-level ac voltage needed by the utility [18].

[**Stefan Achleitner, *et al.*, 2014**] proposed sun-oriented irradiance expectation framework utilizing remote sensors networks for estimating sun-based irradiance for sunlight-based force age with legitimate gauging the changeability effects could be decreased in this way helping the administration of shortening and subordinate generators, expanding inverter productivity with MPPT [19].

[**Christian Viehweger, et al., 2014**] proposed a simulation of concealing impacts and extension of the influenced modules. Shadowing impact on a PV module can happen because of reasons like snow, ice, foliage, soil, and numerous other reasons. In the proposed simulation technique, by applying the newton Raphson method or regular falsie method or bisection method we can apply sensors to the boards, estimating voltage, current, temperature and light [20].

[**Heesung Park, et al., 2014**] proposed the strategy for greatest force point following and force limit technique to improve the sun-powered cluster controller for low earth circle satellites. In customary/common electrical force frameworks of satellites, a steady voltage (CV) mode controls the battery from overcharging accordingly lessening the need of sun-based energy. The proposed strategy shows an improvement of force transformation effectiveness compared to the traditional technique [21].

Some of the researchers used the hardware, software, and technology to enhance the renewable energy as follows.

Authors	Types of renewable energy	Hardware/ Software used	Technology
Hassan Moghbelli et al., Jinn-Chang Wu et al.	Solar Energy Generation Resources	Battery Energy Stockpiling Framework & Seven-Level Inverter	Fixed Tilt panels, Unused Roof Tops of Residential Premises, Trackers
Hida Y. et al., Kalkhambkar V. et al.	Microgrid	HOMER, MATLAB	Simulation, Fuzzy system & Neural Network
Aggarwal A. et al.	Solar Thermal Hybrid System	Thermoelectric material	FGM water tube, Heating Rod, Water tank.

Canada, A.H. et al., Sharaf, A.M. et al.	Solar Energy Resource, Hybrid Solar Wind Power Generation	IRES Method, TMY & RMY data sets, RET Screen Software, Wavelet theory, Wavelet decomposition, V-I Tracer, Simulation, Neural Networks	RMSE, NASA Satellite & BSRN, GIS Technique, Support vector machine & Euclidean method, Electric Vehicle Battery, Wavelet Variability model, Genetic Algorithms, AC-DC busbar, LPP, PLL Control
Hassan Moghbelli et al.	Wind Energy Power Generation	Simulation	BESS, STATCOM

3.2.1.1 Common Issues of Long-Term Sustainable Solar Power Generation

- Different researchers suggested executing PV energy age on the rooftop of the house.
- Scientists investigated the achievability of PV cells, power devices and miniature turbines for upgrading the age of environmentally friendly power and simultaneously lessening CO_2, SO_x, and NO_x discharges [1].
- Researchers of renewable energy decided on the effect of various climate conditions to further develop solar-based azimuth following accuracy and decrease the energy utilization of global positioning systems [2].
- The authors investigated the scientific technique for Optimal estimating of PV-Battery blend in a grid-associated framework [3].
- Researchers investigated the design of programmed age control systems utilizing a fuzzy neural control framework to diminish reliance on diesel motors and the arrangement

approach brought through simulations utilizing MATLAB [4].
- The authors showed the utilization of solar-based paints instead of PV cells for producing renewable energy.
- Researchers exhibited a critical improvement in youngsters' schooling, improvement in their way of life, and a decrease in crime percentage after the development of electrification [5].
- Researchers exhibited the crossover approach for PV power estimation by consolidating wind turbine framework with firefly (FF) and fuzzy framework ARTMAP which upgraded sunlight-based PV age estimate [9].
- Scientists analyzed the impact of direct and indirect skylight on a photovoltaic earth surface by the hour [13].
- Researchers investigated the relationship between the force age proficiency and the varying levels of direct and diffuse solar irradiance across time [16].
- Analysts broke down half and half the photovoltaic energy component power framework to successfully force the executives [17].
- To estimate stack and lessen CO_2 emission, experts have proposed the use of the BESS Battery energy stockpiling system on the demand side of a PV system [20].
- Experts investigated how to reduce the price of solar-based energy while increasing its efficiency by using a combination of a programmed worldwide positioning framework, maximum power point tracking, and logical battery stockpiling charging technology [18].
- In this presentation, the authors showed how deploying portable clusters in photovoltaic converters can increase the output power characteristics [11].

3.2.1.2 Strengths and Weakness Strength

- The establishment of solar boards on unused private premises tackled the energy issue.
- The miniature network was a generally suitable and savvy arrangement in far-off areas [3].
- Worked on the proficiency of CPVGS and hostile to twist limit of the sun-based plate.

- Further develops the force unwavering quality of SPG and diminished force dissemination of the PGS.
- Give adequate power to lighting and water siphoning in country regions [6].
- Photovoltaic power can likewise be utilized to work radio, TV, and cell phone use, which links the towns with the rest of the world.
- Upgrades and huge effect on training, exchange and business, diversion, well-being and so forth [7].
- In the far-off areas with great source speed SWEG undertakings can be monetarily alluring when contrasted with PV power tasks or DG sets to the tune of about 20% is given.
- Wind Turbine has generally higher yearly Capacity Utilization Factor in contrast with sun-oriented photovoltaic or sun-based nuclear energy stations [9].
- Wind entrance further develops the framework voltage profile in the dissemination framework, yet power misfortune increments because of network code prerequisites in huge scope wind farm [11].
- Optimal estimating procedure decreased the misfortunes up to 83% with legitimate measuring of battery applied.
- Wireless sensor networks offered incredible potential as a minimal expense high precision approach for momentary sun-based determining.
- The application of Swinging entryway calculation helps in determining Solar Wind power rises [14].
- Solar paints assimilated low radiations and extremely modest (right around one-third cost of PV cells) and were reasonable in environments.
- Solar home lighting frameworks in rural regions and far-off towns impact battery life for individuals and decrease the cost of lamp oil [15].
- Hybrid innovation utilized in solar breeze power age lessens the working expense.
- Exchanging one switch at high recurrence diminished exchanging power misfortune and further developed the influence of productivity.
- The reliance of episode points of diffuse sun-based irradiance on power age expanded effectiveness by 20% [16].

- A hybrid framework contrasted with fuel and batteries coordinated and PV power frameworks gives continuous excellent force.
- Increased nature of PV productivity by displaying execution of site explicit information.
- Utilize domestic and foreign hardware to discover precise evaluation of huge scope sunlight-based energy.
- Hybrid innovation utilized in solar breeze power age lessens the working expense [18].

Weakness

- High capital expense of PV modules.
- The high cost of energy units and miniature turbines and the reliability of value power from miniature framework was one of the issues [2].
- CPVGS was utilized distinctly in created nations.
- Only 15 minutes ahead of estimates can be produced for sun-powered PV determining [4].
- No streamlining was finished with various sunlight-based PV boards [5].
- Wireless models may not work in the same way for various geological areas.
- Solar Paints come in dim shadings and the existence of painting was little when contrasted with the life of sunlight-based boards [7].
- Large space required.
- Relatively higher introductory capital expense prerequisite.
- The use of a transformer in a DC-DC converter corrupts power proficiency [9].
- The high cost of SVG forestalls the enormous scope of the establishment of SVG.
- CO_2 outflow can't be diminished by introducing BESS [12].
- Trackers were costly.
- Lack of government approaches and impetuses.
- Longer period reference information is needed for more precision [15].
- Low proficiency in thermoelectric material.
- The maintenance cost of models is not considered.
- Regulatory issues like specialized guidelines and sanctioning were significant issues.

- Too numerous specialists to be followed and arranged and observed [16].
- Electricity organizations regularly have a practice of serving for the most part metropolitan populaces through half-breed frameworks and hence might be hesitant to help country charge programs [18].
- Technical matters like troubles with introducing electric gear in conventional structures.
- The cost of off-lattice power is for the most part higher than the expense of traditional force, and off-framework jolt is additionally presently not a substitute for network access for a lot of power that numerous mechanical exercises require [20].
- Due to the unreasonable climate (monetarily helpless and peripheral market), private financial backers are hesitant to put resources into the country zip area [21].

3.3 Conclusion

The current situation favors different cities benefiting from the supply of power communicated by harnessing locally available renewable energy. In order to design and carry out such intervention drives, it is necessary to have concrete evidence of niches that would directly benefit from such a strategy, as well as to investigate and evaluate all viable options for power supply. The availability of environmentally friendly power assets needs to be evaluated, and energy overviews need to be led, in both grid-connected and off-grid communities. Sustainable power-based systems will be planned, and their yearly power delivery will be evaluated, based on the findings of energy studies and other optional information collections. While considering the systems envisioned in the investigation, we will also analyze the role of the customer's willingness and ability to pay for environmentally friendly power-based power supply. Considering the foregoing research on renewable energy, one can surely see the significance of generating electrical power from renewable energy in the next years and how difficulties like low effectiveness and partial shading sway should be addressed. Information gleaned from this research can be used to create a power-appropriation model that accounts for the future's expected variations in energy demand and load profiles. The research will also be helpful for places that, due to a lack of information on solar energy, are unable to produce sustainable energy. The local community and younger students will be reached through a variety of awareness-raising activities designed to promote the transition to renewable energy sources and encourage the conservation of nonrenewable energy. To meet the rising energy needs of cities while simultaneously lowering their carbon footprint, the widespread adoption of renewable energy sources is a crucial future sustainable energy alternative. Renewable energy sources will continue to improve in terms of efficiency, usability, cost-effectiveness, accessibility, and sustainability as technology develops further.

References

1. Canada, A.H., "Solar voltaic generation power plants: 1000 MW to 10 kW photovoltaic plant design and application guide for the Pacific Northwest", *Northcon 95. I EEE Technical Applications Conference and Workshops Northcon95, 10-12 Oct. 1995, Portland, OR, USA.*
2. Suzuki, R., Kawamura, H., Yamanaka, S., Kawamura, H., Ohno, H., Naito, K., "Loss factors affecting power generation efficiency of a PV module", *Photovoltaic Specialists Conference, 2002. Conference Record of the Twenty-Ninth IEEE*, pp. 1557-1560, 19-24 May 2002.
3. Zhenhua Jiang, "Power management of hybrid photovoltaic - fuel cell power systems", *Power Engineering Society General Meeting, 2006. IEEE, Montreal, Que.*
4. Hassan Moghbelli, Robert Vartanian, "Implemenatation of the Movable Photovoltaic Array to Increase Output Power of the Solar Cells", *Proceedings of the International Conference on Renewable Energy for Developing Countries, 25-33, August 2006.*
5. Ladner-Garcia, H.P., O'Neill-Carrillo, E., "Determining realistic photovoltaic generation targets in an isolated power system", *Power & Energy Society General Meeting, 2009. PES '09. IEEE* pp. 1 - 5, 26-30 July 2009, Calgary, AB.
6. Dai, Qinghui, Chen, Jun, "Improving the efficiency of solar photovoltaic power generation in several important ways", *Technology and Innovation Conference 2009 (ITIC 2009), International*, 1-3, 12-14 Oct. 2009, Xian, China.
7. Hou, H.J., Yang, Y.P., Cui, Y.H., Gao, S., and Pan, Y.X., "Assessment of concentrating solar power prospect in China", *Sustainable Power Generation and Supply, 2009. SUPERGEN '09. International Conference on*, vol., no., pp.1, 5, 6-7 April 2009.
8. McIntyre, K.C., Clancey-Rivera, C., Tobin, M.C., Erickson, M.D., Zhang, X., "The feasibility of an environmentally friendly microgrid", *North American Power Symposium (NAPS), 2010*, pp 1-6, 26-28 Sept. 2010, Arlington, TX.

9. XiaoliXu, Qiushuang Liu, YunboZuo, "A Study on All-Weather Flexible Auto-tracking Control Strategy of High-Efficiency Solar Concentrating Photovoltaic Power Generation System", *Intelligent Systems (GCIS), 2010 Second WRI Global Congress,* Volume 2, pp. 375-378,16-17 Dec. 2010, Wuhan.
10. Yahyavi, M., Vaziri, M., Vadhva, S., "Solar energy in a volume and efficiency in solar power generation", *Information Reuse and Integration (IRI), 2010 IEEE International Conference,* pp. 394-399, 4-6 Aug. 2010, Las Vegas, NV.
11. Hida, Y., Ito, Y., Yokoyama, R., Iba, K., "A study of optimal capacity of PV and battery energy storage system distributed in demand side", *Universities Power Engineering Conference (UPEC), 2010 45th International,* pp. 1-5, Aug. 31 2010-Sept. 3 2010, Cardiff, Wales.
12. Huang Chao-Hui, Zhang Xiao-Gang, Yao Xin-li, "A method for micro-gird automatic generation control", *Electronics, Communications and Control (ICECC), 2011 International Conference,* pp. 1224-1227, 9-11 Sept. 2011, Zhejiang.
13. Aggarwal, A., Gupta, A., Kotwal, S., "Sustainable powering: Configuration of renewable sources exclusively for powering a house", *Computer and Communication Technology (ICCCT), 2011 2nd International Conference,* pp. 286-291, 15-17 Sept. 2011, Allahabad.
14. Sharaf, A.M., Aktaibi, A.A., "A novel hybrid facts based renewable energy scheme for village electricity", *Innovations in Intelligent Systems and Applications (INISTA), 2012 International Symposium on ,* vol., no., pp. 1,5, 2-4 July 2012.
15. Tuyishimire, B., McCann, R., Bute, J., "Evaluation of a Kalman predictor approach in forecasting PV solar power generation", *Power Electronics for Distributed Generation Systems (PEDG), 2013 4th IEEE International Symposium,* pp. 1-6, 8-11 July 2013, Rogers, AR.
16. Juuso, E.K., Yebra, L.J., "Optimisation of solar energy collection with smart adaptive control", *Industrial Electronics Society, IECON 2013 - 39th Annual Conference of the IEEE,* pp. 7938-7943, 10-13 Nov. 2013, Vienna.
17. Kalkhambkar, V., Kumar, R., Bhakar, R., "Optimal sizing of PV-battery for loss reduction and intermittency mitigation", *Recent Advances and Innovations in Engineering (ICRAIE), 2014,* pp. 1-6, 9-11 May 2014, Jaipur.
18. Jinn-Chang Wu, Chia-Wei Chou, "A Solar Power Generation System With a Seven-Level Inverter", *Power Electronics, IEEE Transactions* Volume 29, Issue 7, pp. 3454 – 3462, 17 February 2014.
19. Achleitner, S., Kamthe, A., Tao Liu, Cerpa, A.E., "SIPS: Solar Irradiance Prediction System", *Information Processing in Sensor Networks, IPSN-14 Proceedings of the 13th International Symposium,* pp. 225-236, 15-17 April 2014, Berlin.

20. Viehweger, C., Hartmann, B., Keutel, T., Kanoun, O., "Simulation of Shading Effects on the power output of solar modules for enhanced efficiency in photovoltaic energy generation", *Instrumentation and Measurement Technology Conference (I2MTC) Proceedings, 2014 IEEE International*, pp. 610-613, 12-15 May 2014, Montevideo.
21. Heesung Park, Hanju Cha, "The design of a solar array regulator with power limitation function", *Industrial Technology (ICIT), 2014 IEEE International Conference*, pp 766-770, Feb. 26 2014-March 1, 2014, Busan.

9. XiaoliXu, Qiushuang Liu, YunboZuo, "A Study on All-Weather Flexible Auto-tracking Control Strategy of High-Efficiency Solar Concentrating Photovoltaic Power Generation System", *Intelligent Systems (GCIS), 2010 Second WRI Global Congress*, Volume 2, pp. 375-378,16-17 Dec. 2010, Wuhan.
10. Yahyavi, M., Vaziri, M., Vadhva, S., "Solar energy in a volume and efficiency in solar power generation", *Information Reuse and Integration (IRI), 2010 IEEE International Conference*, pp. 394-399, 4-6 Aug. 2010, Las Vegas, NV.
11. Hida, Y., Ito, Y., Yokoyama, R., Iba, K., "A study of optimal capacity of PV and battery energy storage system distributed in demand side", *Universities Power Engineering Conference (UPEC), 2010 45th International*, pp. 1-5, Aug. 31 2010-Sept. 3 2010, Cardiff, Wales.
12. Huang Chao-Hui, Zhang Xiao-Gang, Yao Xin-li, "A method for micro-gird automatic generation control", *Electronics, Communications and Control (ICECC), 2011 International Conference*, pp. 1224-1227, 9-11 Sept. 2011, Zhejiang.
13. Aggarwal, A., Gupta, A., Kotwal, S., "Sustainable powering: Configuration of renewable sources exclusively for powering a house", *Computer and Communication Technology (ICCCT), 2011 2nd International Conference*, pp. 286-291, 15-17 Sept. 2011, Allahabad.
14. Sharaf, A.M., Aktaibi, A.A., "A novel hybrid facts based renewable energy scheme for village electricity", *Innovations in Intelligent Systems and Applications (INISTA), 2012 International Symposium on*, vol., no., pp. 1,5, 2-4 July 2012.
15. Tuyishimire, B., McCann, R., Bute, J., "Evaluation of a Kalman predictor approach in forecasting PV solar power generation", *Power Electronics for Distributed Generation Systems (PEDG), 2013 4th IEEE International Symposium*, pp. 1-6, 8-11 July 2013, Rogers, AR.
16. Juuso, E.K., Yebra, L.J., "Optimisation of solar energy collection with smart adaptive control", *Industrial Electronics Society, IECON 2013 - 39th Annual Conference of the IEEE*, pp. 7938-7943, 10-13 Nov. 2013, Vienna.
17. Kalkhambkar, V., Kumar, R., Bhakar, R., "Optimal sizing of PV-battery for loss reduction and intermittency mitigation", *Recent Advances and Innovations in Engineering (ICRAIE), 2014*, pp. 1-6, 9-11 May 2014, Jaipur.
18. Jinn-Chang Wu, Chia-Wei Chou, "A Solar Power Generation System With a Seven-Level Inverter", *Power Electronics, IEEE Transactions* Volume 29, Issue 7, pp. 3454 – 3462, 17 February 2014.
19. Achleitner, S., Kamthe, A., Tao Liu, Cerpa, A.E., "SIPS: Solar Irradiance Prediction System", *Information Processing in Sensor Networks, IPSN-14 Proceedings of the 13th International Symposium*, pp. 225-236, 15-17 April 2014, Berlin.

20. Viehweger, C., Hartmann, B., Keutel, T., Kanoun, O., "Simulation of Shading Effects on the power output of solar modules for enhanced efficiency in photovoltaic energy generation", *Instrumentation and Measurement Technology Conference (I2MTC) Proceedings, 2014 IEEE International*, pp. 610-613, 12-15 May 2014, Montevideo.
21. Heesung Park, Hanju Cha, "The design of a solar array regulator with power limitation function", *Industrial Technology (ICIT), 2014 IEEE International Conference*, pp 766-770, Feb. 26 2014-March 1, 2014, Busan.

4

Approaches for District-Scale Urban Energy Quantification and Rooftop Solar Photovoltaic Energy Potential Assessment

Faiz Ahmed Chundeli[1]* and Adinarayanane Ramamurthy[2]

[1]Department of Architecture, School of Planning and Architecture, Vijayawada, Andhra Pradesh, India
[2]Department of Planning, School of Planning and Architecture, Vijayawada, Andhra Pradesh, India

Abstract

With increasing population and urbanization, cities are expected to draw energy demands from diverse sources, especially local on-site generation. For energy-efficient urban design and planning, understanding the energy demand and consumption at the city scale is essential. Therefore, the use of alternate means of energy generation and advanced technologies such as photovoltaic (PV) is pertinent. This chapter uses a simplified methodology for the quantification of consumption and a possible supply of energy demands of an urban district by classifying them into archetypes, a bottom-up urban energy modelling technique. Climate data, land use, building use, height, floor area, occupancy, the assumption on HVAC-dependent areas, wall-to-window ratio (WWR), U-value of materials, roof area, surface areas respective to cardinal orientations, and annual solar radiation are used for classification of archetypes, energy quantification and PV potentials. Based on this classification, 917 buildings in the study area are classified into 14 archetypes with eight different residential and two different commercial, mixed-use and institutional archetypes, respectively. MIT Design Advisor rapid building energy simulation tool is used to quantify annual cooling and lighting energy demands of classified archetypes. The weighted average net energy demands of the archetypes are then scaled to the selected study area using GIS. A district-scale 3D urban energy map and solar PV potential map of the case area are generated for simplified interpretations. Based on the potential PV analysis,

---

Corresponding author: faizahmed@spav.ac.in; faizahmed.arch@gmail.com

Deepak Kumar (ed.) Urban Energy Systems: Modeling and Simulation for Smart Cities, (47–64) © 2023 Scrivener Publishing LLC

it is found that, on average, 0.58-0.52 kWh/m² of electricity can be generated in the study area.

Keywords: Archetypes, urban energy, neighborhood energy efficiency, photovoltaic, district energy

4.1 Introduction

Urban energy systems that consume less energy and emit less carbon are the need of the hour. Further, the causal relationship between urbanisation and energy use demands efficient urban energy systems. The demand for energy for household consumption, commercial utilisation, etc., has increased tremendously with increasing urbanisation across the globe, especially in South Asian countries. The urban population constitutes about 50% of the world's population and 75% of the global energy demand. With the increasing population and urbanisation worldwide, there is an urgent need to revamp urban energy generation. The demand for urban energy shall continue to rise in the coming decades to accommodate the growing urban population. Therefore, any potential steps to transform the urban energy system would be a step towards sustainability [23]. Although methods of building energy simulation (BES) have been well established, urban building energy simulation (UBES) is still a nascent field [20].

Furthermore, energy quantification and performance at the district scale have still not been thoroughly explored in the Indian context. A comprehensive review of district-scale urban energy systems modelling is carried out by several researchers, such as [3, 7, 13, 15, 16, 20, 24]. Based on the literature, urban or district energy demands are estimated using top-down or bottom-up modelling approaches. A bottom-up model is a data-driven intense modelling approach, while a top-down is a policy-driven one. Until recently, the top-down modelling approach has been in practice, and the bottom-up approach has evolved recently [15]. Currently, bottom-up modelling is becoming popular because of its statistical and analytical approach. Cities are expected to adopt energy-efficient and renewable energy sources to meet the demands.

Further, recent urban development policies emphasise the need for such measures to attain sustainable development goals. Urban energy demands are now met by adopting new technologies, such as PV, for energy generation [7, 8, 19, 21, 30]. Further, the literature suggests that PV has been one of the renewable energy technologies that have gained most acceptance over recent years [5, 11]. At the end of 2011, the global use of PV as a renewable source had reached 65,000 GW [7, 11].

The central focus of India's development is to achieve the energy goals of meeting the electricity and fuel demand, thereby improving the infrastructure for better living conditions. Since 2000, energy demand and consumption have increased by a hundred times and are predicted to rise exponentially [27]. Coal is by far the most crucial fuel in the energy scenario in the Indian case, and the recent climate pledge underlined the country's commitment to alternative energy sources such as solar and wind. According to Tamil Nadu transmission corporation data, Tamil Nadu (TN) met an all-time peak power demand of 14,538 MW in February 2016, whereas in 2015, it was around 12,000MW–13,000MW [2]. This chapter presents district-scale urban energy quantification of the Kannadasan Nagar area (T. Nagar), Chennai (TN), India. A simplified archetype classification is based on parameters like building use, building footprint, number of floors, occupancy, building orientation, etc. MIT Design Advisor, a simplified whole-building energy simulation tool, has been used for simulating the archetypes for estimating the annual energy demand in terms of heating, cooling, and lighting [25, 26]. Rooftop PV potential analyses were carried out for a possible renewable generation. Using Geographic Information System (GIS), the quantified archetype energy was extrapolated for the entire district. This chapter intends to explain a simplified process of urban energy assessment using 3D geospatial technology for informed decision-making. The chapter also presents possible renewable energy generation through PV for the same case area.

4.2 District-Scale Urban Energy Modelling

Literature suggests several models and approaches for quantifying urban energy through computer simulations [3]. Many approaches are available based on data requirement, accuracy, applicability, etc., addressing single or multiple issues related to urban energy simulation. This section of the chapter presents the modelling approaches adopted in this research, including a "bottom-up" approach based on archetype classification and a renewable energy modelling approach.

4.2.1 "Bottom-Up" Modelling Approach – Archetype

Several models are available for quantifying and assessing building energy demand. A bottom-up energy simulation modelling approach is commonly used for scaling up from building to urban. There are two ways to model the bottom-up approach. The first method is based on processing the individual buildings statistically and analytically to measure the building energy usage via linear regression [15]. The second method is by using building physics. In the building physics modelling approach, archetypes of the case study area are evolved. These archetypes are developed based on the building's physics and material properties, such as window-to-wall ratio (WWR), building age, dwelling type, building use, local climatic zones, U-value of the material, etc. [15, 16]. These archetypes are information-rich 3D models that include microclimate information, thermal properties of the building, including occupant behaviour, and building physics parameters [17, 22, 28]. The archetypes are modelled in the whole building energy simulation tool, energy consumptions are quantified, and the results of archetypes are extrapolated on neighbourhood or urban blocks' energy demand assessment and performance [18].

4.2.2 The Renewable Energy Modelling Approach

To reduce transmission loss, gird congestion, etc., cities are expected to meet their demands through local energy generation through renewable energy generation integration [19, 23, 24]. Further, identifying local energy generation potentials through renewable sources is one of the feasible ways to meet the urban energy demand. Solar energy is a renewable source that needs to be tapped for long-term sustainability. Urban areas have the potential for harvesting solar energy to generate electricity through rooftops and building surfaces. For assessing the solar potentials of a given building, understanding the total solar irradiations falling on rooftops and the effective utility space available for installing PV panels are crucial [5].

4.2.3 Urban Microclimate

Building performance and energy demands for heating, cooling, and lighting are significantly impacted by the local microclimate of a place [14, 17, 28]. Varied temperature profiles are observed within the city regions. Therefore, urban energy assessment and quantification must consider the variations in local micro-climate in terms of temperature, wind velocity, etc.; these variations need to be integrated. Further, buildings cannot be evaluated in isolation for their performance as they are part of a whole urban or neighbourhood block. In addition, solar rations, wind velocity, and urban heat islands must be considered in urban energy simulations [22, 28].

District-Scale Urban Energy Assessment 51

Figure 4.1 Kannadasan Nagar, Chennai - case area location map (Source: Prepared by the authors.).

4.3 Evaluation of Energy Performance – The Case in Chennai

4.3.1 Profile of the Case Area

In this chapter, Kannadasan Nagar, one of the planning units in Chennai city corporation is identified for study [6]. Kannadasan Nagar is part of Theagaraya Nagar (T. Nagar), a commercial hotspot of Chennai and one of India's most prominent shopping districts. 0.69 sq. km is the total planning unit area of Kannadasan Nagar. In 1980, Chennai Metropolitan Development Authority (CMDA) prepared a Detailed Development Plan (DDP) for the study area. Based on the primary data collected by the authors, as of 2013, the total population of the study was 77,280, with more than 900 buildings within the study area. Figure 4.2 shows the footprint of the study area along with building uses.

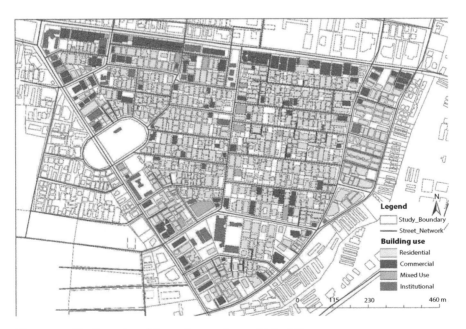

Figure 4.2 The footprint of the study area along with building uses.

4.3.2 Data Model and Construction Techniques

3D Geographic Information Systems (GIS) are used in this study for mapping the building footprint, heights, and use of the entire study area. Further, GIS is used to generate information-rich 3D models of the study area for further analysis. For the digitisation of the building footprint, Google imagery was used, and the same was validated through the primary survey. Further, the compatibility building use assessment with the proposed land-use map of Chennai 2026 was carried out. The proposed master plan of 2016 was used to extract regulatory zoning information such as floor space index (FSI), building setback, the maximum allowable height of buildings, etc. 2D ArcGIS is used to create a geospatial database of the study area using the primary data collected from the survey. Further, 3D ArcScene is used for generating a 3D rich information model of the study area integrating the zoning regulation [1]. Based on the mapping, it was found that the total residential, commercial, mixed-use, institutional, parks, and playgrounds account for about 53%, 0.17%, 18%, 8%, and 2%, respectively [6]. Further, of the total study area of about 0.69 sq. km, about 19% of the land is utilised for road infrastructure. For validation of the primary data collected from the study area, all the plot-level zoning regulatory parameters were verified through the secondary sources available with Chennai Metropolitan Development Authority (CMDA). Since there is no secondary information for cross-validation of information about buildings, the entire buildings in the study area were enumerated. The primary survey mapped all the buildings and the number of floors and uses. The same is tabulated and presented in Table 4.1. The information thus collected is used for generating a 3D database of the study area.

4.3.3 Archetype Classification

Fourteen archetypes are generated by analysing 917 surveyed buildings in the study area. The parameters for generating archetypes are building use, the number of floors, occupancy rate, window-to-wall ratio (WWR), and orientation. Eight residential archetypes start from ground structure (RB-T1) to Apartments – ground+ four structures (RBA1-T1) and two different commercials, mixed-use and institutional archetypes, respectively, as shown in Figure 4.3, and the details can be found in Table 4.2.

Table 4.1 Floor-wise area statement of the study area, 2013.

Floors	R-count	R-area m²	C-count	C-area m²	I-count	I-area m²	Total count	Total area m²
I	713	188580	169	65742	35	16866	917	271188
II	675	184844	153	60664	25	13343	853	258852
III	279	105411	66	32850	8	5847	353	144107
IV	136	58836	33	20461	2	1613	171	80910
V	36	18962	22	14705	1	1070	59	34738
VI	11	6724	12	9565	0	0	23	16288
VII	4	3648	4	4240	0	0	8	7888
VIII	3	3018	4	4240	0	0	7	7258
IX	3	3018	2	1974	0	0	5	4992
X	1	1360	0	0	0	0	1	1360
		574401		214440		38740		827580
		69%		26%		5%		

R – Residential, C – Commercial, I – Institutional.
Source: *Primary survey by the author.*

District-Scale Urban Energy Assessment 55

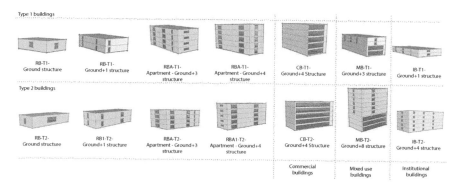

Figure 4.3 Indicative representations of archetypes.

4.3.4 Energy Quantification

The MIT Design Advisor, a simplified whole-building energy simulation tool, is used to quantify the total annual energy consumption in terms of cooling, heating, and light loads [25, 26]. This simple-to-use tool is developed to enable rapid building energy quantification. The tool provides quick, visual comparisons which are needed for energy quantification. An annual energy simulation can quantify heating, cooling, and lighting loads with input data and parameters such as temperature, humidity, orientation, building dimensions, occupancy, equipment used, lighting requirements, WWR, etc. Annual building energy loads for all 14 archetypes are quantified and tabled (Table 4.2) using the following characteristics.

    i.   Building use: Residential, Commercial, Mixed-use, and Institutional buildings
   ii.   Orientation of the buildings: North-south or east-west
  iii.   Density (based on area and household): Low-rise residential, high-rise residential, office buildings, and classrooms
  iv.   Number of floors: Ground, ground+1, ground+2, ground+3 structures, etc.
   v.   Occupancy rate: Residential: 0.03 – 1; Commercial & mixed use; 0.1; Institutional: 0.25
  vi.   Lighting requirements: Rough work – 300 lux; office work – 500 lux
 vii.   Equipment usage (w/m^2): Residential: 5-10 (light-medium); Commercial -15 (medium-high); Mixed-use: 5-10 (light-medium); Institutional: 1-5 (light)

Table 4.2 Archetype building energy quantification.

Archetype	Building configuration (m)	No. of floors	Total area (m²)	Orientation	Air change (litres/sec/person)	Airflow rate (mouthfuls per hour)	Occupancy rate (people/m²)	Annual cooling load (kWh/m²)	Annual lighting load (kWh/m²)	Total annual energy consumption (kWh/m²)	Roof area (m²)	Available roof area (RUF: 70%) m²	Solar energy considering losses (kWh)/m²	Annual solar energy generated (kWh)
Residential buildings														
RB-T1	11 X 15	11	165	E-W	2.5	0.11	0.03	70	10	157	165	115.5	0.62	72
RB-T2	21 X 8	1	168	N-S	2.5	0.11	0.029	90	50	77	168	117.6	0.62	73
RB1-T1	12 X 21	2	504	E-W	2.5	0.11	0.075	100	25	350	252	176.4	0.61	109
RB1-T2	22 X 9	2	396	N-S	2.5	0.15	0.1	110	10	204	198	138.6	0.62	86
RBA-T1	13 X 30	4	1560	E-W	2.5	0.11	0.075	100	30	520	390	273	0.61	169
RBA-T2	30 X 20	4	2400	N-S	2.5	0.08	0.1	130	30	708	600	420	0.61	260
RBA1-T1	9 X 33	5	1485	E-W	2.5	0.02	0.025	95	95	510	297	207.9	0.62	129
RBA1-T2	33 X 11	5	1815	N-S	2.5	0.6	1	270	25	1455	363	254.1	0.62	157
Commercial buildings														
CB-T1	16 X 32	5	2560	E-W	5	0.09	0.1	125	20	1315	512	358.4	0.62	222
CB-T2	34 X 16	5	2720	N-S	5	0.09	0.1	135	15	635	544	380.8	0.62	235
Mixed-use building														
MB-T1	17 X 39	4	2652	E-W	2.5	0.06	0.1	190	45	782	663	464.1	0.62	288
MB-T2	27 X 18	9	7290	N-S	2.5	0.03	0.1	280	30	1095	486	340.2	0.62	211
Institutional buildings														
IB-T1	32 X 84	2	5376	E-W	5	0.56	0.25	130	60	364	2688	1881.6	0.62	1166
IB-T2	50 X 22	5	5500	N-S	5	0.23	0.25	180	70	515	1100	770	0.58	447

Constants for archetype energy quantification:
Max. Temp – 41C; Min. temp- 20C; Max. relative humidity – 90%
Lighting levels: Rough work – 300 lux; Office work – 500 lux
Equipment: Light-5; Medium-10; Medium-High-15 w/m²
WWR: Residential -25%; Commercial-90%; Institutional-30%

Constants for solar PV potential quantification:
Annual solar irradiation on the titled surface: 5.51kWh
The conversion efficiency of panels/Solar panel yield: 15%
Performance Ratio, the coefficient for loss: 0.75

viii. Air Change Rate and Fresh Air Rate (litres/sec*person): Residential & mixed use: 2.5; Commercial & institutional: 5
ix. Window-to-wall ratio (WWR): Residential: 25%; Commercial: 90%; Institutional: 30%

4.3.5 Analysis of the Archetype Energy Quantification

The total of 917 buildings in the study area is classified into 14 archetypes, and there are eight different residential archetypes and two different commercial, mixed-use, and institutional archetypes, respectively, as shown in Figure 4.3. Building energy quantification of each archetype is carried out considering the following parameters: building orientation, floor details and building heights, occupancy rate, lighting requirement, equipment used, air change and fresh air rates, and wall-to-window ratio (WWR). After the archetype classification, the energy consumption for each building was quantified, and it was observed that, on average, buildings oriented to the east-west axis consumed more energy than buildings oriented along the north-south axis. As tabulated, the equipment and lighting requirements also showed significant variation in energy consumption levels by the archetypes (Table 4.2). It is found that the per capita energy consumed by the residential archetype varied between 0.4-0.7 kWh/m^2.

Further, it was observed from the study that commercial archetypes consumed an average of 0.4 kWh/m^2. Similar results are observed for mixed-use archetypes; however, it was established that institutional building archetypes consume about 0.1 kWh/m^2. Energy quantification of all the 14 archetypes, along with all the parameters assumed for calculation, is presented in Table 4.2.

4.3.6 Solar PV Potential Calculation

Provision for rooftop solar PV installation is feasible for most buildings surveyed. Effective utilisation of the rooftops, excluding overhead tanks and staircase headroom, which are about 6 m^2 and 30 m^2, are eliminated for calculations. Further, it may be noted that rooftop solar PV installations depend on the effective useable roof space in any building. Based on the primary survey, 70% of the total terrace area has been assumed for installations. Therefore, of 100 m^2 of any terrace area, 70 m^2 is assumed to be used for solar PV installation. The same has been used as space utilisation quantification. All the buildings in the study area are assessed for potential solar PV installation using the archetypes and solar exposure analysis, i.e., solar insolation analysis. Overall energy loss while conversion of solar to electricity is accounted for arriving at the net energy generation from the study area. The following formula is used to calculate potential annual solar energy; the same is tabulated in Table 4.2.

> *Potential Solar Energy E (kWh) = Total solar panel Area (A in m^2) * Solar panel yield (r in %) * Annual average irradiation on tilted panels (H, shading is not included) * Performance ratio (PR, Performance ratio, the coefficient for losses (range between 0.9 and 0.5, the default value of 0.75, Considering 15% losses)*

4.3.7 Analysis of Solar PV Potential

The analysis of the available roof area of individual archetypes and the solar insolation shows that, on average, 10-45% of the energy consumed can be compensated using PV technology. On average, 0.58-0.62 kWh/m^2 can be generated from a unit roof area of the case study. Solar PV potentials of all the 14 archetypes, along with all the parameters assumed for calculation, are presented in Table 4.2. Not all the building owners would agree to PV installations, but for research purposes, it is assumed that all the buildings in the case area would go for PV systems installation. It should be noted that by law, all institutional buildings across the country must meet 30% of their energy consumption by solar, and the government of India provides up to 50% subsidy for installing PV for residential buildings [27].

4.3.8 Scaling of Archetype Building Energy to District-Scale Urban Energy

At the urban scale, energy utilization analysis and the map will help decision-makers optimize the planning and management of urban energy systems [9, 10, 29]. GIS allows the generation of urban energy maps by combining spatially geo-referenced data with simulated archetype results [4, 12]. In this chapter, the representative archetype results based on residential, commercial, mixed-use, and institutional use are scaled to individual building uses. A weighted average statistical technique minimizes the deviation in individual energy consumption while scaling up. Figure 4.4 shows the annual energy demand for individual buildings for meeting cooling and lighting loads. The 3D urban energy map presented here helps understand the demand created by individual buildings across the case area. 3D urban models have proved effective in many cases, including in urban planning processes and decision-making [1, 10]. Simplified

District-Scale Urban Energy Assessment 59

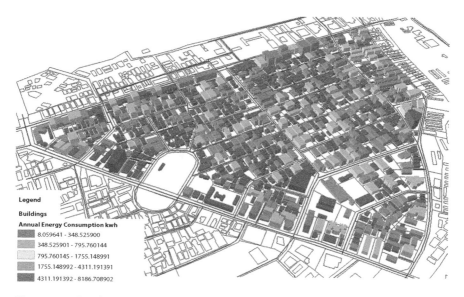

Figure 4.4 3D urban energy map of the case area.

3D information-rich models such as solar PV potentials can aid decision-makers. Not only are informed decisions made possible using this model, but they also aid in sensitising the users on per capita energy usage. Apart from creating awareness of energy utilisation, the model shall be helpful for urban planners and architects. The results show that the annual energy demand for the entire study area is slightly over 403 MW.

4.3.9 Scaling of Archetype PV Potential to District-Scale PV Potential

Similar to scaling building energy consumption to district energy consumption, the PV potentials of individual archetypes are scaled to the entire roof area of the case by taking a 0.7 utilization factor [5]. The total annual solar energy that can be harnessed through PV is about 118 MW, catering to about 29% of the total energy demand of the case area. However, it is not possible to use 100% of the roof area to install PV systems, but the result projects the opportunity for local renewable energy generation. Figure 4.5 shows the building of the case area and PV potential roof area in 3D.

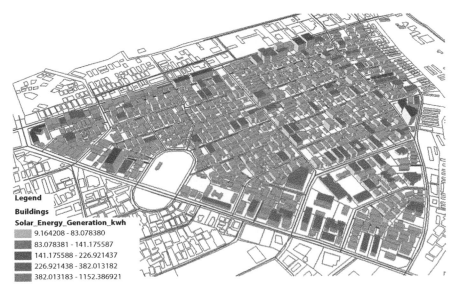

Figure 4.5 PV potential of the case is in 3D.

4.4 Discussions and Conclusions

4.4.1 Discussion

Literature from across the globe indicates the inclination of researchers towards developing advanced technologies for harvesting renewable energy sources for attaining sustainable development goals. Such research requires a multidisciplinary approach combining urban planning and design with urban energy science. Though many studies are carried out at the building level integrating new technologies, neighborhood or urban district studies are still merging [20]. The neighborhood or district-level energy demand for cooling, heating, and lighting can be assessed and optimised in a simplified manner through 3D urban models. Further, 3D urban models can improve informed decision-making while planning and designing urban areas. Urban energy maps may also support the decision-makers in verifying the buildings in terms of their energy usage in near real-time. It will also help evaluate the performance of buildings in terms of their building orientation, massing, openings, occupancy, and other building characteristics.

Many studies have shown that 3D urban models can be used to sensitize citizens on various urban planning and environmental and climate issues.

Further, the 3D urban models can be used for analysing the performance of the neighborhood or urban blocks, in addition to exploring possibilities of energy-saving initiatives. Though the applications of 3D urban models are handy, the challenge is simplifying the data and modelling techniques. Data availability and simplified way for communicating the simulated results are re-emerging at district scale urban energy quantification, which was prevalent at building level energy quantification. With an explicit database generated for quantification of district-scale urban building energy simulation, further 3D urban analysis can be performed, such as processing data for accuracy, visualisation and real-time dynamic energy simulation, dissemination of urban energy data with citizens, etc. [1, 4, 12]. A dynamic 3D urban energy building model integrating building characteristics and urban physics would be a way forward in this field.

With changing lifestyles, modernisation, and access to numerous electrical appliances, energy demands are bound to increase. Thereby, the per capita increase in energy demand needs to be addressed sustainably. Therefore, the use of renewable energy sources shall continue to be investigated to meet the supply gap and increase the energy generation potential onsite. Hence, in this study, the use of solar PV and the positive impact that it would create is demonstrated. While developed countries are on the path of tapping renewable energy generation sources, developing countries, especially India, where urbanisation is at its peak, require more attention from researchers [27]. With expanding tier 1 and tier 2 cities, architects, urban planners, and decision-makers must consider how renewable energy sources can be tapped effectively. Therefore, planners and urban energy managers must be equipped with simplified assessment tools for informed decision-making.

4.5 Conclusions

In this chapter, an easy methodology for district-scale urban energy quantification uses bottom-up urban energy modelling techniques via archetype. Building physics characteristics like floor area, window-to-wall ratio (WWR), type of dwelling, building use, climatic zone, U-Value, etc., are used to generate archetypes. Rapid building energy simulation tool MIT Design Advisor is used to quantify total annual energy consumption in terms of cooling and lighting load for the classified archetypes of the case area Kanndasan Nagar (T. Nagar), Chennai, India. The simulated results of archetypes were then scaled to the entire buildings of the case area, and urban energy consumption and performance maps were generated.

The results found that, on average, 0.1-0.7 kWh/m² of energy demand per unit area was consumed. In this chapter, a demonstration of simplified methodology for district-scale urban energy quantification and rooftop solar photovoltaic potential is presented by taking a pilot study in the city of Chennai. However, the entire city can be assessed for energy generation potentials using the same methodology. The methods thus demonstrated shall be helpful for architects, urban planners, and decision-makers to quickly assess and quantify energy generation potentials in a simplified format. Also, a detailed urban rooftop solar PV potential of the case area was carried out as an alternative local source of energy generation. Based on the findings, about 62.33 kWh of energy can be generated in a 100 m² effective terrace utilization area, and about 29% of the annual energy demand of the study area can be met through solar PV installation.

Further, the chapter demonstrated the application of 3D urban models and 3D GIS for visualisation and informed decision-making. The results of such 3D models may be more accurate considering the extracted building profile and context. In addition, it is to be noted that further simplification of 3D data collection and 3D modelling is the need of the hour.

References

1. Ahmed, F. C., & Sekar, S. P. (2015). Using Three-Dimensional Volumetric Analysis in Everyday Urban Planning Processes. *Applied Spatial Analysis and Policy*, 8(4), 393–408. https://doi.org/10.1007/s12061-014-9122-2
2. Ahmed, C. (2017). Smart Geo-Spatial Analytical Tool for Urban Development and Management. *Proceedings of the Special Collection on e-government Innovations in India - ICEGOV '17*, 76-81. http://dx.doi.org/10.1145/3055219.3055238
3. Allegrini, J., Orehounig, K., Mavromatidis, G., Ruesch, F., Dorer, V., & Evins, R. (2015). A review of modelling approaches and tools for the simulation of district-scale energy systems. *Renewable and Sustainable Energy Reviews*, 52, 1391–1404. https://doi.org/http://dx.doi.org/10.1016/j.rser.2015.07.123
4. Ascione, F., De Masi, R. F., de Rossi, F., Fistola, R., Sasso, M., & Vanoli, G. P. (2013). Analysis and diagnosis of the energy performance of buildings and districts: Methodology, validation and development of Urban Energy Maps. *Cities*, 35, 270–283. https://doi.org/http://dx.doi.org/10.1016/j.cities.2013.04.012
5. Bergamasco, L., & Asinari, P. (2011). Scalable methodology for the photovoltaic solar energy potential assessment based on available roof surface area: Further improvements by ortho-image analysis and application to

Turin (Italy). *Solar Energy*, 85(11), 2741–2756. https://doi.org/http://dx.doi.org/10.1016/j.solener.2011.08.010
6. CMDA. (2008). Chennai II Master plan. Chennai. Retrieved from http://www.cmdachennai.gov.in/smp_main.html
7. Connolly, D., Lund, H., Mathiesen, B. V, & Leahy, M. (2010). A review of computer tools for analysing the integration of renewable energy into various energy systems. *Applied Energy*, 87(4), 1059–1082. https://doi.org/http://dx.doi.org/10.1016/j.apenergy.2009.09.026
8. Crawley, D. B., Hand, J. W., Kummert, M., & Griffith, B. T. (2008). Contrasting the capabilities of building energy performance simulation programs. *Building and Environment*, 43(4), 661–673. https://doi.org/http://dx.doi.org/10.1016/j.buildenv.2006.10.027
9. Dall'O', G., Galante, A., & Torri, M. (2012). A methodology for the energy performance classification of residential building stock on an urban scale. *Energy and Buildings*, 48, 211–219. https://doi.org/http://dx.doi.org/10.1016/j.enbuild.2012.01.034
10. Deevi, B., & Chundeli, F. (2020). Quantitative outdoor thermal comfort assessment of street: A case in a warm and humid climate of India. *Urban Climate*, 34, 100718. doi: 10.1016/j.uclim.2020.100718
11. Gautam, B. R., Li, F., & Ru, G. (2015). Assessment of urban rooftop solar photovoltaic potential to solve power shortage problem in Nepal. *Energy and Buildings*, 86, 735–744. https://doi.org/http://dx.doi.org/10.1016/j.enbuild.2014.10.038
12. Howard, B., Parshall, L., Thompson, J., Hammer, S., Dickinson, J., & Modi, V. (2012). Spatial distribution of urban building energy consumption by end-use. *Energy and Buildings*, 45, 141–151. https://doi.org/http://dx.doi.org/10.1016/j.enbuild.2011.10.061
13. Jebaraj, S., & Iniyan, S. (2006). A review of energy models. *Renewable and Sustainable Energy Reviews*, 10(4), 281–311. https://doi.org/http://dx.doi.org/10.1016/j.rser.2004.09.004
14. Kalz, D. E., Pfafferott, J., & Herkel, S. (2010). Building signatures: A holistic approach to the evaluation of heating and cooling concepts. *Building and Environment*, 45(3), 632–646. https://doi.org/http://dx.doi.org/10.1016/j.buildenv.2009.07.016
15. Kavgic, M., Mavrogianni, A., Mumovic, D., Summerfield, A., Stevanovic, Z., & Djurovic-Petrovic, M. (2010). A review of bottom-up building stock models for energy consumption in the residential sector. *Building and Environment*, 45(7), 1683–1697. https://doi.org/http://dx.doi.org/10.1016/j.buildenv.2010.01.021
16. Keirstead, J., Jennings, M., & Sivakumar, A. (2012). A review of urban energy system models: Approaches, challenges and opportunities. *Renewable and Sustainable Energy Reviews*, 16(6), 3847–3866. https://doi.org/http://dx.doi.org/10.1016/j.rser.2012.02.047

17. Kwok, S. S. K., Yuen, R. K. K., & Lee, E. W. M. (2011). An intelligent approach to assessing the effect of building occupancy on building cooling load prediction. *Building and Environment*, 46(8), 1681–1690. https://doi.org/http://dx.doi.org/10.1016/j.buildenv.2011.02.008
18. Li, Z., Quan, S. J., & Yang, P. P.-J. (2016). Energy performance simulation for planning a low carbon neighborhood urban district: A case study in the city of Macau. *Habitat International*, 53, 206–214. https://doi.org/http://dx.doi.org/10.1016/j.habitatint.2015.11.010
19. Lund, P. D., Mikkola, J., & Ypyä, J. (2015). Smart energy system design for large clean power schemes in urban areas. *Journal of Cleaner Production*, 103, 437–445. https://doi.org/http://dx.doi.org/10.1016/j.jclepro.2014.06.005
20. Reinhart, C. F., & Cerezo Davila, C. (2016). Urban building energy modeling – A review of a nascent field. *Building and Environment*, 97, 196–202. https://doi.org/http://dx.doi.org/10.1016/j.buildenv.2015.12.001
21. Robinson, D. (2011). *Computer modelling for sustainable urban design: physical principles, methods and applications*. London: Routledge.
22. Santamouris, M. (2001). *Energy and climate in the urban built environment*. London: Earthscan.
23. Shimoda, Y., Fujii, T., Morikawa, T., & Mizuno, M. (2004). Residential end-use energy simulation at city scale. *Building and Environment*, 39(8), 959–967. https://doi.org/http://dx.doi.org/10.1016/j.buildenv.2004.01.020
24. Swan, L. G., & Ugursal, V. I. (2009). Modeling of end-use energy consumption in the residential sector: A review of modeling techniques. *Renewable and Sustainable Energy Reviews*, 13(8), 1819–1835. https://doi.org/http://dx.doi.org/10.1016/j.rser.2008.09.033
25. Urban Bryan, J. L. R. G. (2006). The MIT Design Advisor: A fast, simple building design tool. In IBPSA.
26. Urban Bryan, J. L. R. G. (2007). A Simplified Rapid Energy Model and Interface for Nontechnical Users (pp. 1–10). Retrieved from http://designadvisor.mit.edu/design/
27. Vimala, M. (2016). Energy consumption in India - recent trends. *Asia Pacific Journal of Research* Vol: I. Issue XXXVI.
28. Wong, N. H., Jusuf, S. K., & Tan, C. L. (2011). Integrated urban microclimate assessment method as a sustainable urban development and urban design tool. *Landscape and Urban Planning*, 100(4), 386–389. https://doi.org/http://dx.doi.org/10.1016/j.landurbplan.2011.02.012
29. Yamaguchi, Y., Shimoda, Y., & Mizuno, M. (2007). Proposal of a modeling approach considering urban form for evaluation of city level energy management. *Energy and Buildings*, 39(5), 580–592. https://doi.org/http://dx.doi.org/10.1016/j.enbuild.2006.09.011
30. Zhao, F, Martinaz-Moyano I.J, & Augenbroe, G. (2011). Agent-based modeling of commercial building stocks for policy support (pp. 2385–2392).

5

Energy Consumption in Urban India: Usage and Ignorance

Rajnish Ratna[1]* and Vikas Chaudhary[2]†

[1]Gedu College of Business Studies, Royal University of Bhutan, Gedu, Bhutan
[2]Manav Rachna University, Faridabad, India

Abstract

Energy consumption and human-caused carbon dioxide emissions rank India as the world's third-highest emitter. For India to achieve its climate targets, it must make quick changes to its urban energy infrastructure. However, there is no comprehensive baseline urban energy-use dataset for all urban districts in India that is comparable to national totals and that integrates social, economic, and infrastructure factors. Because of its negative consequences on the global environment and its danger to long-term sustainability, the current pattern of energy usage (particularly based on fossil-based fuels) raises severe concerns. It is our moral and legal responsibility to develop an energy consumption pattern that maximizes efficiency, safeguards the environment, and keeps and improves the vitality of our regional economy. This article centres on urban energy, namely renewable and sustainable options. Energy consumption across the globe has been on the rise for decades. Getting people to reduce the squandering of energy is one way to cut down on overall usage. Learning current consumer behaviours, attitudes in terms of motivation, barriers towards conservation, and readiness to invest in understanding and conserving energy are only a few of the elements that must be taken into account when designing appropriate motivational technology.

Keywords: Energy consumption, urban India, sustainability, conservation, energy audit

---

*Corresponding author: rajnish.gcbs@rub.edu.bt; rajnish.ratna@gmail.com
†Corresponding author: rchdry@yahoo.co.in

5.1 Background

When it comes to human-caused carbon emissions and global energy consumption, India ranks third. India must make significant changes to its urban energy infrastructure to achieve its climate goals as it rapidly urbanizes [1]. Still, no standard data set for urban energy usage covers all of India's metro districts consistent with national statistics and combines the social, economic, and infrastructure aspects relevant to such changes [2–4]. India's recent energy progress has been tremendous, yet the country still faces several obstacles. The spread of COVID-19 has been troublesome [5–8]. Nevertheless, India has made significant advancements in recent years to increase the availability of power, particularly for the poor, to promote the widespread adoption of energy-efficient Lighting systems, and pushed for a vast increase in renewable energy, primarily solar power. As a result, improvements in the standard of living for Indian citizens are visible as seen in Table 5.1. Increased climate change considerations have prevented more extensive use of energy technology options. As a result, it is challenging to provide India with enough energy, of the right quality, in various forms and at prices that people can afford [9, 10].

This study presents research that compared the consequences of different energy technology options under different growth rates and nuclear and renewable energy usage scenarios to determine their potential influence in the future. The study used various models to analyze potential energy sources within the context of user-specified limits such as fuel supply, environmental regulations, etc. After the research period, in the year 2035, it is anticipated that the electricity demand will be 50 040 MW/Yr [11].

In this thorough analysis, coal has remained the primary fuel for power generation. Coal imports were already signed before the research period but grew even more during it. Around 11,000,000,000,000,000 tonnes of carbon dioxide will be emitted between now and 2035 if the baseline

Table 5.1 Various energy demands projected for 2035.

Thermal	428170 MW/Yr
Hydro	40320 MW/Yr
Nuclear	1400 MW/Yr
Wind	18000 MW/Yr

scenario comes to fruition. When the share of nuclear energy climbed to 9% from 3% in the baseline scenario, carbon dioxide emissions decreased by 10% in the high nuclear capacity scenario.

5.2 Introduction

India, the world's largest democracy, has a population of around 1.05 billion people and is seeing fast economic expansion, mainly owing to policy measures taken by the government over the previous decade, which have resulted in quicker GDP growth. As a result, some forecasting firms estimate that India's economy will continue to rise over the next 30-40 years. Coal, oil, and solid biomass supply 80% of the world's energy needs, although energy consumption has grown since 2000 [12–14]. As a result, India has one of the lowest global energy consumption rates and carbon dioxide emissions per person. This trend holds for several other critical indicators, including the number of cars on the road and the amount of steel and cement produced. If India is to expand at the present possible rate, it needs a reliable supply of energy, especially power and crude oil, at rates that are competitive on a global scale. The energy plan's primary goal is to continuously meet the energy needs of all sectors. India's annual per capita electricity usage of roughly 750 kWh (including captive power generation) is extremely low compared to other developed countries. In addition, many rural areas lack modern conveniences like running water and electricity. All energy sources, from exploration/mining to transit to generation to final usage, come with environmental costs and risks. According to studies by the Global Commission on Climate Change, global warming is due to the increased atmospheric carbon dioxide that humans have created. India must plan its energy infrastructure in light of its resource profile, technological foundation, and human resource capacity. As a result, India would have to rely more and more on imports from a small number of far-off, often politically unstable regions for its oil and gas needs. For this reason, it is essential to assess the current energy situation for immediate and long-term purposes. Despite significant progress, the current state of the country's power supply is marked by chronic shortages, unreliability, and high prices for industrial consumers. The situation with petroleum products is also causing some anxiety.

There are concerns regarding India's energy supplies because the country relies on oil imports to 70%. Even if electricity can be generated domestically, its availability is limited by factors such as the accessibility of coal, the viability of developing hydropower, and the potential for increasing nuclear power. New technologies for the efficient use of energy, and electricity, in particular, take time to develop, making scenario building a useful tool for identifying priority areas and kicking off R&D. Energy security, R&D, addressing environmental concerns, energy saving, etc., are all areas where progress could benefit from more defined goals.

In 2020, natural gas and cutting-edge renewable energy sources started making headway and were least impacted by the COVID-19 pandemic. In addition, solar photovoltaic (PV) has seen a spectacular rise; the resource potential is huge, and government support and technological price cutting have made it the low-priced option for new power generation.

An analysis of India's energy demand growth rates was conducted, focusing on electricity, and methods were developed to satisfy the expected demand under various scenarios, all with the overarching goal of this study to evaluate India's existing and future power demands, and to analyze future electricity industry developments.

5.3 Energy Outlook for India

India's energy sector is becoming increasingly significant. It is the fifth-largest economy globally and is a major energy importer. As a result, the country's primary energy source, coal, would undoubtedly increase imports [15, 16].

Increasing energy consumption directly results from the region's high economic growth, including India. Over the past half-century, India has increased its overall energy consumption, mostly coming from commercial rather than noncommercial energy sources. According to primary commercial energy production trends, coal has been the most plentiful commercial energy source over the past five decades. Domestic production and supply increases in the petroleum and natural gas industries. However, fuelwood, agricultural residue, and animal manure, among other noncommercial energy sources, still provide a significant portion of energy needs, especially in rural households. The use of alternatives is gradually replacing the rural sector's reliance on traditional energy sources, commercial sources that are both more reliable and efficient. Despite increasing resources and expanding energy supplies, India still has severe energy shortages. As a result, more and more energy is being imported to meet needs. Following is an overview of the leading commercial energy sources and power generation methods [10, 17].

A. Coal

India is among the three leading coal producers worldwide. Coal is primarily utilized in the energy sector. In addition, coal is used in producing steel, cement, fertilizers, chemicals, paper, and many more products on a large and small scale. India's three most significant states regarding coal production are Bihar, Madhya Pradesh, and West Bengal. In January 2005, the Geological Survey of India assessed that the country had 247.85 BT in total coal reserves, which included all deposits up to a depth of 1,200 metres. This total consists of 92.96 TB of verified data, 117.1 TB of indicated data, and 32.97 TB of inferred data. The country's coal reserves were estimated to be 35.95 BT in January 2005. The sole basis for these reserves is geological research. Roughly 382 million metric tonnes of coal were produced in the United States in 2004-2005, while a net 25.3 million metric tonnes were brought in from abroad. In addition, roughly 24.8 million metric tonnes of lignite were produced. As a result of the great distances that separate some of India's most critical consuming centres from their nearest sources of resources, India's transportation network is under tremendous strain. Since coal plays a significant role in the Indian energy environment, it is crucial to build a comprehensive energy security plan for the country.

B. Oil and Gas

The rapid growth of India's economy has increased the demand for hydrocarbons. The amount of crude oil extracted in 2004-2005 was 33.98 million tonnes, while the amount of natural gas extracted was 31.77 billion cubic metres. The majority (75%) of India's crude oil supply in 2004/05 came from imports, totalling 95.86 million tonnes, to satisfy rising demand. The government has offered the industry to participate in the private sector to meet expanding demand. Private entrepreneurs have contributed substantially to exploring new oil and gas deposits. However, because of the magnitude of its crude oil imports, a country is particularly vulnerable to supply shocks, which could have far-reaching effects on the economy.

C. Power Generation

When it comes to a country's economic growth, electricity is one of the most crucial inputs in terms of infrastructure. The installed generating capacity in India increased to about 118,419 MW during March 2005. This includes all environmentally friendly sources, like solar (2,980 MW), wind (2,980 MW), hydro (30,936 MW), and nuclear (3,850 MW) (831 MW). Starting in October 2007 [2], nuclear power accounted for 4.7% of all electricity generated. Thermal power plants account for roughly 68%, hydro for 26%, nuclear for 3%, wind for 2%, and other renewables for 1%. Presently, utilities generate around 587 BU annually in terms of power. Additional power is produced in-house to the tune of 71.58 BU. The electricity industry relies heavily on coal. From 1986–1987 to 2003–2004, India's annual per capita power consumption rose from 191 kWh to 592 kWh. The average annual per-person power use is estimated to be 192.33 kWh, but this number varies widely across regions. The industrial sector constitutes a significant portion of India's total power consumption in several regions. Many companies, like those that make aluminium, cement, iron and steel, paper, fertilizer, textiles, and sugar, have built their captive power plants so they don't have to rely on the power grid. These captive plants are meant to serve as a backup to the purchased electricity and to be used in the event of a power outage.

D. Hydro Energy
India can generate 150,000 MW of electricity from hydropower. However, there is an uneven distribution of hydroelectric potential, with 83% concentrated in the north and northeast. Only roughly 31 GWe had been developed or were in development as of March 2005. The last decade has seen a decrease in the expansion of hydroelectric power, which was supposed to account for 40% of the generation capacity. The necessity for significant financial investments, the length of time between the completion of feasibility studies and the start of operations, worries regarding the influence on the environment and public opinion, and other factors are among the reasons. The Ministry of Power has recently taken several measures to hasten the addition of capacity to hydroelectric projects to use the country's hydropower potential fully. This new hydropower programme will allow the construction of 162 new hydroelectric facilities throughout 16 states, producing 47,930 MW.

E. Non-conventional Renewable Energy
Our nation's non-conventional renewable energy sources have a capacity of approximately one hundred gigawatts of electricity. Wind, small hydro, and biomass power all offer significant untapped potential, while solar photovoltaics (PV), solar thermal, and waste-to-energy conversion are also crucial components. Particularly in outlying locations, there will be a growing demand for all these supplies in the years to come. In the intermediate term, we aim to guarantee at least 10 GWe of renewable energy by 2012. Recent years have seen enormous gains in wind power's installed capacity due to all the progress made in the sector. However, the generation capacity provided by wind turbines is poor.

F. Nuclear Energy
India's nuclear power plants are now Pressurized Heavy Water Reactors (PHWR), which have been and are being developed and constructed exclusively in India. The amount of power that can be generated from the United States' local uranium resource using such reactors is around 330 GWe-yr, which is about the same as having 10 GWe of PHWRs installed and running at 80% of their lifetime capacity for 40 years. Fast Breeder Reactors (FBR) can generate an estimated 42,200 GWe annually by recycling uranium. Considering that 60% of the heavy metal is utilized, this is uranium. It is also studied that if the FBR is run for 100 years with an 80% lifetime capacity factor, its potential generation is similar to an installed capacity of about 530 GWe (*International Journal of Civil and Environmental Engineering* 1:3 2009 116). Thorium deposits can produce around 150,000 GWe-yr, which, if recycled multiple times through the appropriate reactor systems, can satisfy India's energy requirements for a long time. The Department of Atomic Energy has mapped out a three-phase nuclear power strategy to make the most available resources. Firstly by 2020, it is anticipated that roughly 20 GWe of nuclear-generating capacity will have been installed.

The second phase of the nuclear power programme calls for constructing a series of fast breeder reactors, which would increase the inventory of fissile materials, the rate at which they are produced, and the power they generate. The construction of FBRs will thus become central to India's nuclear power initiative. Third, fast or thermal critical reactors, or accelerator-driven sub-critical reactors, should use the country's abundant thorium resources.

5.4 Power Demand and Resources in India

Energy consumption and demand in India have increased at one of the fastest rates in the world in recent years. This increase can be attributed to the country's rapidly growing population and expanding economy. Between 1981 and 2001, businesses experienced a 6% rise in the quantity of primary energy they used. However, despite the overall rise in power demand, the amount of energy consumed in Power Capital is still relatively low compared to that of other developing countries. This is because both non-renewable and renewable forms of energy are readily available in abundant quantities in India. Commercial energy is dominated by coal, petroleum, and gas. Until the late 1980s, India's energy policy was built on the country's ability to rely on domestic energy production. As a result, coal was the primary source of energy use. In this section, we'll go through some of the various home-grown options for power. With current production levels, the maximum amount of indigenous coal available can grow by three times. The estimated 12 BT of hydrocarbon reserves could produce around 1,200 EJ of energy. Furthermore, energy security over the long term can be ensured by deploying nuclear technology for power generation due to the abundance of uranium and thorium compared to fossil fuel supplies. To meet the ever-increasing need for energy, it is essential to conduct activities such as discovering new fossil fuel reserves, cost-effectively importing energy, optimizing hydroelectric power generation and expanding the use of nuclear and other non-fossil energy sources. In addition, if measures are taken to limit emissions of greenhouse gases, nuclear fuel supplies may help close the gap between supply and demand for energy.

The government of India has established energy and electricity requirements estimates to formulate an Integrated Energy Policy. Economic growth, population growth, the rate at which non-renewable energy sources are phased out in favour of commercial energy, gains in energy conservation and efficiency, and sociological and lifestyle changes all play a role in these forecasts.

According to the Census of India in 2011, India is divided into 640 districts, distributed among 27 states and eight union territories. There might be both towns and villages included within one district. According to the Census of India results in 2011, the term "urban areas" refers to census towns and statute towns combined. Statutory towns are administrative divisions that are designated as having urban character. Census towns are more miniature than statutory towns.

Approximately 31% of India's total population resides in urban regions, according to the Census of India conducted in 2011. Therefore, even though the energy consumption in urban regions was the primary focus of our research, we also evaluated the energy consumption in rural areas to ensure that the sum of the energy consumption in all districts was comparable to the total energy consumption in the country.

End-use sectors of energy comprised residential, commercial, and industrial manufacturing, as well as transportation, agricultural, and other commercial and public sectors. Coal, various types of petroleum fuel, electricity, and firewood were all considered forms of end-use energy (only in the residential sector). The data came from the 2011 Census of India, the 68th Round of the National Sample Survey, and the Annual Survey of Industry in 2012. The amount of fuel used for vehicular transportation on roads was estimated based on the number of cars and the distance they travelled. We adopted a top-down approach to reduce the government of India's reported national energy use to the district level by factoring in the number of people employed in each sector. Commercial enterprises, farming, and manufacturing processes relying on petroleum fuel were among the many areas where this concept was implemented. When developing top-down estimates of energy consumption across industries and fuel types, the National Energy Statistical Yearbook and the Indian Petroleum and Natural Gas Statistics were the primary data sources. Publicly available social-demographic, economic, infrastructure, and urban form characteristics were combined with modelled data on energy use by end-use industries in all urban regions. The 2011 Census of India provided most of the data for these variables.

5.5 Energy and Environment

Air quality

Energy consumption is the primary contributor to India's air pollution, which has become one of the country's most serious environmental challenges in recent years. Around a million people's lives were cut short in 2019 due to exposure to air pollution in their homes and communities. Only in China is the death rate from preventable causes higher than in India. There are 124 "non-attainment cities" in 24 of India's 36 states and union territories, according to the Central Pollution Control Board (CPCB) and the National Green Tribunal. As a result of the fact that not all of India's approximately 500 cities have air quality monitoring stations, the number of 124 cities may be underestimated. This is a significant amount, but it may be lower.

Despite more research, monitoring, and corrective action being focused on urban areas, rural India still has poor air quality due to the widespread use of biomass for cooking. Therefore, air pollution has become a critical issue for the well-being of the Indian populace. The government of India acknowledged the severity of the issue and, in 2019, launched the National Clean Air Programme (NCAP), which has the objective of lowering the concentrations of PM2.5 and coarse particulate matter (PM10) by 20-30% by 2024 in comparison to the levels that existed in 2017. This is the first time that a deadline for meeting an air quality improvement objective has been established, and while it won't ensure that cities will meet NAAQS standards, it is a significant step in the right direction.

Land

Hydropower projects posed the biggest threat to land use in India's energy sector. However, there are still some challenging legacy concerns, such as resettling and rehabilitating displaced people because of dam building, even though much of this hydroelectric capacity was installed years ago. In addition, the new push in India for utility-scale solar has also brought up issues with land acquisition and utilization. An analysis by the National Institute of Solar Energy found that only 3% of India's wastelands would be sufficient to support 750 gigawatts (GW) of solar PV.

The high population density across the country and the propensity for land holdings to be divided add to the challenges. For example, in 2015, the typical agricultural owner had 1 hectare of land. However, several initiatives have been taken to address these challenges and make acquiring land for solar power infrastructure development easier. For example, when the SECI partners with state governments to create solar parks, the states take care of land acquisition.

Carbon Emissions

Even though India's per capita emissions are low compared to the rest of the world, the country is responsible for 14% of the growth in global energy-sector emissions since 1990. The three main sectors responsible for the meteoric rise in India's energy-related CO_2 emissions since 1990 are electricity generation, industry, and transportation. Electricity generation contributed roughly twice as much to the increase in 1990 emissions as did all other industries. Coal is responsible for 70% of India's CO_2 emissions from the energy industry yet supplies just 45% of India's primary energy requirements. Compared to the global average of 510 g CO_2/kWh, India's power industry emits 725 g CO_2/kWh, demonstrating the disproportionately enormous contribution of inefficient coal-fired generation.

The industrial sector is the second-greatest emitter after the electricity generation sector. Although it only accounts for about 6% of world steel production, India's iron and steel industry is the country's most energy-intensive and, thus, one of the biggest contributors to pollution. Cement manufacturing in India accounts for about 8% of worldwide output and is the second most polluting industrial sector. Fuel consumption for freight transportation by road in India has significantly increased since 2000, making it the second highest globally, behind only China. Trucks account for more than 45% of India's total road transportation emissions. Also quadrupling since 2000 are emissions from passenger vehicles on India's roads. Since two- and three-wheeled vehicles make up a disproportionately large portion of India's fleet (about four times as many as passenger cars), this helps to explain why only 18% of the country's total transport emissions come from cars.

Water
Currently, the energy sector collects around 30 billion cubic meters (bcm) of water from sources and uses nearly six bcm of that water. For cooling and ash disposal, coal-fired power plants use the great bulk of the water extracted for use in the energy industry. Only 5% of India's total water withdrawals and 2% of consumption goes toward the energy industry, but access to adequate water is vital to the country's energy security.

5.6 Sustainable Development Goals (SDGs) for Indian Electricity Sector

The power sector in India has seen extraordinary growth during the past ten years. The grid was able to be synchronized for the first time in 2013 when it became one of the largest in the world. With the help of a massive electrification effort, nearly every family worldwide can access electrical power. Over the past decade, over two hundred billion dollars have been invested in expanding energy transmission and distribution networks. As a direct result, India is now the world's fourth-largest energy market, following the United States, China, and the European Union. Increased access to electricity has allowed for the rise of a middle class that can afford modern conveniences and the upkeep of a digitally oriented way of life. As a result, the increase in India's electricity demand has become an increasingly popular indicator of the country's overall economic performance. On the supply side, India has emerged as a global leader in the solar business thanks to the spectacular growth of utility-scale photovoltaic (PV) systems, which competitive auctions have supported. Currently, six states in India get between 10% and 30% of their electricity from variable renewable sources.

Access to energy, energy security, and energy sustainability are only a few policy goals India will need to pursue concurrently to support the country's expanding population and economy. To do this successfully, it is necessary to approach the formulation of energy policies and the planning of energy policies with an eye toward the implications that various policy choices will have on the system. In India, for instance, the demand for transportation may be affected by changes in the capacity of power plants, the designs of refineries, the infrastructure for the supply of natural gas, the emissions of greenhouse gases, and the quality of the air. Moreover, India's decisions have far-reaching implications for the overall course of global tendencies. In this overview, we highlight these interrelationships through scenario planning and analysis.

Because various ministries and organizations formulate them, India's policies and goals regarding the energy industry are frequently inconsistent and sector-specific. Nevertheless, all these policies and goals are connected in some way to the overarching objective of delivering energy services that are reliable, inexpensive, and environmentally friendly. According to its Nationally Determined Contribution (NDC), India plans to lower the carbon intensity of its economy by 33–35% from 2005 levels by 2030 and to increase the share of non-fossil fuels in its capacity for power generation to 40% by the same year. In addition, the Indian Prime Minister has identified seven areas of focus for developing India's energy industry. These include moving toward a "gas-based economy," using fossil fuels more cleanly, using more biofuels, rapidly scaling up renewable energy sources, prioritizing electric mobility, shifting toward emerging fuels like hydrogen, and innovating energy systems digitally. By 2030, we need to have increased our renewable power capacity by 450 gigawatts, increased the proportion of natural gas in our primary energy mix to 15%, sold 30% more electric vehicles relative to passenger vehicles, and blended 20% more biofuels into our gasoline. In the 2020s, there is also an effort to reduce crude oil imports and end coal imports, increase energy efficiency in all sectors, provide affordable housing, electrify trains, and more. These goals are outlined in the Sustainable Development Goals (SDGs).

Access to economic, efficient, and cutting-edge energy services
Access to the electric grid: Over the past few years, India has made significant strides toward expanding access to electricity through the Saubhagya Scheme. More than 99% of Indian homes had access to the electrical grid in 2019, according to official government statistics. However, there remain persistent problems with the reliability and quality of power for consumers who do not have their own homes connected to the grid. According to the findings of several studies, fewer than 80% of customers at institutional establishments, 65% of customers at small enterprises, and in 2018, 50% of farms with customers were connected to the grid [18, 19].

It is anticipated that between 2021 and 2030, an annual average of around $35 billion will be spent on the installation of transmission and distribution lines and maintenance. In addition, subsidy programmes will enable low-income households to raise their electricity use, which is a substantial rise from current spending levels. Electricity demand in homes is expected to increase from 2019's average of around 1,000 kWh per year to an average in 2030 of about 2,000 kWh per year due to increased household incomes and enhanced central grid reliability.

Availability of clean cooking options
Access to safe cooking methods has moved forward at a much more glacial pace than access to electricity has, partly because of the prohibitive costs involved and the lack of available supplies.

The government has made tremendous efforts through the PMUY programme to increase citizens' access to LPG. While nearly all Indian homes (97.5% as of 2015) have access to LPG, we still expect over 650 million people (or just under half of the country's population) to utilize biomass for most of their cooking needs [20–22]. Even though LPG is now widely available to homes, this remains a problem (and so are counted as not having access). More than 90% of those who lack access to something live in rural areas, making this a problem that is especially prevalent there.

Energy and pollution in the air
The attainment of several Sustainable Development Goals (SDGs) is contingent on improvements in air quality. Air pollution can be reduced in several ways: by expanding renewable energy's share, which would cut down on pollution in the power sector; by boosting energy efficiency, which would cut down on pollution across the board in the energy economy; and by assuring access to clean energy, which would cut down on pollution in homes; and by increasing access to sustainable transportation, which would improve air quality in cities. In addition, direct co-benefits for reducing greenhouse gas emissions can be gained from implementing policies to minimize air pollution.

Emissions caused by coal
As part of its Nationally Determined Contribution (NDC), India has committed to increasing the use of non-fossil fuels in electricity generation from 2005 to 40% by 2030. The Paris Agreement binds members to this objective. As a result, around 60% of the world's total energy capacity will be in India by 2030, and the country's emissions intensity of GDP will be 40% lower than in 2005. These two broad objectives are both met by the Stepping Towards Enhancing Policy Structures (STEPS).

78   URBAN ENERGY SYSTEMS

5.7   Results

A. Scenario in Urban Areas Focusing Businesses
In this section, we go over the results of the MESSAGE model research on the current state of the Indian electricity grid. Throughout the analysis period, overall electricity use was multiplied. Throughout this analysis, coal has been and continues to be the most widely used fossil fuel. Without any intervention, nucleases' share of the pie grows more extensive at the end of the research period. In a status quo scenario, nuclear power can only produce 20 GW. Throughout this analysis, hydropower's impact gradually lessened. In this scenario, wind power technology's contribution expanded dramatically throughout the research period up to its maximum capacity of 45 GW because it is readily available at a reasonable cost. Despite coal's continued dominance as a fuel source during the study period, the study indicated that by 2035, about 70% of the coal used in the United States would have to be imported due to insufficient domestic production. During the study's first phase, the oil contributions are constrained because that resource must be shared with other commercial energy sectors, such as the transportation industry.

Coal, natural gas, oil, hydro, nuclear, and renewables all contribute to the final percentage shares of 81%, 15%, 8%, 3%, and 4%, respectively. The model estimates that only 8% of total electricity generation will come from hydropower, although 84 GW of hydropower capacity will be used. Between 2015 and 2030, several new nuclear power facilities are scheduled to go online. In addition, wind and hydroelectric power facilities are being built at a steady rate during the duration of the study.

5.8   Conclusions

Throughout this study, coal has been and will continue to be the most widely used fuel for generating energy. Substituting nuclear power for coal reduces emissions of greenhouse gases by about 10%. Coal is only partially replaced by renewable energy, yet that still results in a 4% decrease in carbon dioxide emissions. By increasing the capacity of nuclear power plants and renewable energy sources, as well as making full use of hydropower, more hybrid scenarios can be built, reducing the need for coal. Carbon dioxide emission regulations can be used as restraints to examine trends further.

References

1. Ajay Singh Nagpure, Mark Reiner, & Anu Ramaswami. (2018, February 13). Resource requirements of inclusive urban development in India: insights from ten cities. *Environmental Research*. https://doi.org/10.1088/1748-9326/aaa4fc
2. Anu Ramaswami, & Tanya Heikkila. (2012, December 3). A Social-Ecological-Infrastructural Systems Framework for Interdisciplinary Study of Sustainable City Systems. https://doi.org/10.1111/j.1530-9290.2012.00566.x
3. Apoorva Pandey, & Chandra Venkataraman. (n.d.). Estimating emissions from the Indian transport sector with on-road fleet composition and traffic volume. *Atmospheric Environment*, 98(2014), 123-133. https://doi.org/10.1016/j.atmosenv.2014.08.039
4. Awareness and adoption of energy efficiency in Indian homes. (2020, October 1). India Environment Portal. Retrieved September 13, 2022, from http://www.indiaenvironmentportal.org.in/content/468778/awareness-and-adoption-of-energy-efficiency-in-indian-homes/
5. Current Status | Ministry of New and Renewable Energy, Government of India. (n.d.). mnre.gov. Retrieved September 13, 2022, from https://mnre.gov.in/bio-energy/current-status
6. Dayal, & A.E. (2019). *Rural Electrification in India: Customer Behaviour and Demand*. The Rockefeller Foundation. Retrieved September 13, 2022, from https://www.rockefellerfoundation.org/report/rural-electrification-india-customer-behaviour-demand/
7. Examining Effects of the COVID-19 National Lockdown on Ambient Air Quality across Urban India. (2020, June 25). Aerosol and Air Quality Research. Retrieved September 13, 2022, from https://aaqr.org/articles/aaqr-20-05-covid-0256
8. *Final Report on Urban Planning Characteristics to Mitigate Climate Change in Context of Urban Heat Island Effect*. (2017). Bangalore: The Energy and Resources Institute. Retrieved September 13, 2022, from https://www.teriin.org/sites/default/files/2018-03/urba-heat-island-effect-report.pdf
9. Ghani, E., Grover, A., & Kerr, W. (2015, October 30). India's chaotic and messy use of energy. World Bank Blogs. Retrieved September 12, 2022, from https://blogs.worldbank.org/developmenttalk/india-s-chaotic-and-messy-use-energy
10. Why India's CO_2 emissions grew strongly in 2017. (2018, March 28). Carbon Brief. Retrieved September 13, 2022, from https://www.carbonbrief.org/guest-post-why-indias-co2-emissions-grew-strongly-in-2017/
11. *India Energy Policy Review 2020 - Event - IEA*. (2020, January 10). International Energy Agency. Retrieved September 13, 2022, from https://www.iea.org/events/india-energy-policy-review-2020

12. Lalitha, L., & Juneja, N. (2020). SME finance in the wake of the pandemic. *Economic Times* Energyworld. https://energy.economictimes.indiatimes.com/news/renewable/opinion-sme-finance-in-the-wake-of-the-pandemic/79229052
13. Peter Zeniewski, & Siddharth Singh. (n.d.). India Energy Outlook 2021. *IEA 2021*. iea.org/reports/india-energy-outlook-2021
14. Power Sector at a Glance ALL INDIA, Government of India. (n.d.). Ministry of Power. Retrieved September 13, 2022, from https://powermin.gov.in/en/content/power-sector-glance-all-india
15. Report on Performance of State Power Utilities 2018 19. (n.d.). Power Finance Corporation, 2020 (New Delhi).
16. Reports on SDG (n.d.). NITI Aayog. Retrieved September 13, 2022, from https://www.niti.gov.in/reports-sdg
17. Saradhi, I.V., G.G. Pandit, & V.D. Puranik. (n.d.). Energy Supply, Demand and Environmental Analysis. *A Case Study of Indian Energy Scenario, 1:3*(2009), 115. 2009
18. Saxena, V. (n.d.). *Inequalities in accessing LPG and electricity consumption in India: the role of caste, tribe, and religion*. St Andrews Research Repository. Retrieved September 13, 2022, from https://research-repository.st-andrews.ac.uk/handle/10023/11401?show=full
19. *A sustainable recovery plan for the energy sector – Sustainable Recovery – Analysis – IEA*. (n.d.). International Energy Agency. Retrieved September 13, 2022, from https://www.iea.org/reports/sustainable-recovery/a-sustainable-recovery-plan-for-the-energy-sector
20. Tim Hillman, Bruce Janson, & Anu Ramaswami. (2009, December 21). Spatial Allocation of Transportation Greenhouse Gas Emissions at the City Scale. https://doi.org/10.1061/(ASCE)TE.1943-5436.0000136
21. Tong, K., Nagpure,, & A.S. & Ramaswami. (2021, April 12). All urban areas' energy use data across 640 districts in India for the year 2011. *Sci Data 8*(2021), 104. https://doi.org/10.1038/s41597-021-00853-7
22. Census of India (2021). Wikipedia. Retrieved September 13, 2022, from https://en.wikipedia.org/wiki/2021_Census_of_India

6
Solar Energy from the Urban Areas: A New Direction Towards Indian Power Sector

Sonal Jain

School of Social, Financial and Human Sciences, Kalinga Institute of Industrial Technology (KIIT) University, Bhubaneswar, India

Abstract

Future energy prosperity depends on renewable energy. India's cheapest energy source is solar power, which has a huge energy potential and produces no hazardous by-products as a renewable energy source. Solar energy systems are now widely available for business and household usage and require very little maintenance. Tax rebates and credits may make solar energy financially viable. Solar power is increasingly popular in wealthier countries. Modern architects design photovoltaic cells and electronics. The National Solar Mission aims to improve India's energy security and promote ecologically responsible economic growth. It will be India's main contribution to global climate change efforts. The National Solar Mission aspires to make India a global leader in solar energy by promoting its fast spread. The Mission's short-term goal is to expand solar energy consumption nationwide. "India is a tropical nation with lengthy, intense sunlight," states the National Action Plan on Climate Change. Solar power has great potential as a sustainable energy source. Decentralized energy delivery provides local communities with more control over their lives and economies.

Keywords: Solar energy, renewable energy, government, tax rebate

6.1 Introduction

To tackle issues like energy security, climate change, and sustainable development on a global scale, the development of cutting-edge clean energy technology must be rapidly accelerated [1, 2]. It is expected that solar

---

Email: sonalmyid@gmail.com; sonalihzbjain@gmail.com, ORCID-0000-0002-7452-5753

photovoltaic will emerge as an appealing alternative power source in the future, making it a crucial technical choice in realizing the transition to a decarbonized energy supply [3, 4]. The growth of the PV solar grid will take place around the world. Energy demand is a common indicator of economic health used by many economists. The worldwide demand for primary energy sources is predicted to increase by a factor of three in the future decades [5, 6]. According to projections made by the International Energy Agency [7], India would have the largest population on Earth by the year 2050. This rise in overall energy demand is linked to this demographic growth. By 2040–50, the country's energy needs would have increased by more than 200%, according to a Greenpeace analysis that focuses on issues including population growth and exponential GDP growth [8–10]. According to these forecasts, oil demand will rise, as will the demand for natural gas. The country's power grid would have to treble in size to meet this sharp increase in energy consumption. The papers not only show how much energy will be needed in the future, but they also include predictions for renewable energy sources by 2050. In India, PV accounts for 20% to 40% of all renewable energy systems, according to a Greenpeace analysis, which predicts growth in PV installations across the nation [11].

Many factors, such as economic growth, social progress, and serious health issues related to the widespread use of fuel wood, charcoal, etc., contribute to the pressing demand for sufficient resources in developing countries like India. The quantity of greenhouse gas emissions from industrialized economies is often far higher than those from developing ones, according to historical trends. According to Kim, India hopes to increase the use of renewable energy by nearly three times by 2030. This will have a huge effect on the worldwide battle against climate change and the nation's energy supply. "The motivation behind these investments is Prime Minister Modi's dedication to renewable energy, particularly solar energy. The World Bank Group will make every effort to assist India in achieving its lofty goals, particularly about expanding solar energy."

India is the third-largest consumer of energy globally. Twenty percent of the country's electricity comes from green energy. Renewable energy comes in many forms, including biomass, photovoltaics, wind, and water power. India is fortunate to receive sunshine that is at its most powerful. The Indian government decided to exploit these numerous renewable energy sources effectively as a result. To encourage the installation of solar energy systems on rooftops in both rural and urban areas, the Indian government has introduced a variety of solar programs. To encourage individuals to adopt solar panels for electricity generation, the government offers a 30% discount on solar energy installations. Users of solar energy can also reduce their electricity costs by up to 90%. Other advantages, including a 25-year warranty on solar panels, are also being the manufacturer of solar panels. The Ministry of New and Renewable Energy has a facility that may be turned out for hosting conferences, workshops, and other types of events. In India, the use of solar energy is being actively promoted by a variety of organizations, including the National Solar Energy Center, the National Wind Energy Center, the Solar Energy Corporation of India, and other organizations.

6.2 Renewable Energy Chain in India

The rapid growth of India's renewable energy industry has opened up several new opportunities for businesses across the country, from those engaged in the sourcing of raw materials to those engaged in the distribution of finished products and the integration of whole systems. The Indian RE sector offers a wide range of services and products based on wind and solar photovoltaic (PV) technology. Many companies, both new and old, are attempting to capitalize on the growing renewable energy market.

Resources for generating electricity and their by-products, raw materials, R&D during design and integration, engineering and equipment manufacturing, EPC contractor selection, operations and maintenance, stakeholder engagements, electrical utilities, testing, and quality assurance are all part of the renewable energy value chain. Electricity and heat can be generated by the sun. Additionally, the government has begun researching India, which has resulted in a multiplicity of chances for manufacturing and service enterprises throughout the value chain, from raw materials to product manufacture and system integration. Wind and solar photovoltaic (PV) technologies supply a comprehensive selection of services and goods in the Indian RE business. Many companies, both new and old, are attempting to capitalize on the growing renewable energy market.

The renewable energy value chain consists of many interconnected parts, including generation resources, the goods they produce, and the raw materials used in their production; research and development during the design and integration phase; engineering and equipment manufacturing processes; the selection of the most suitable EPC contractor; and finally, operation and maintenance, multi-stakeholder engagements, electrical utilities, testing, and quality assurance. Solar energy may be used to provide both power and heat. Solar photovoltaic (PV) parks on a gigawatt (GW) scale and solar PV rooftop installations on a megawatt (MW) scale are both options that have been investigated in the country. So far, flat plate

84   URBAN ENERGY SYSTEMS

collector technologies have dominated the residential section of the solar thermal technology market. However, to meet the right amount of heat energy at the right moment, commercial and industrial sectors are increasingly using concentrator technologies like parabolic and Arun dishes (with double-axis tracking) and Scheffler dishes (with single-axis tracking).

6.3   Development of Solar Photovoltaic and Solar Thermal Plants

The 2006 Rural Electrification Program was the Indian government's first move toward acknowledging the value of solar energy. It provided instructions for putting off-grid solar applications into action. However, as of 14 February 2012, just 33.8 MW of capacity has been added using this scheme. This mostly refers to solar lamps, solar pumps, lighting systems for homes, streets, and homes, as well as solar lighting systems. India's Semiconductor Policy was first implemented in 2007 as a follow-up measure to support the IT and electronic sectors.

Being a tropical country, India has a lot of potential for solar power. Depending on their location and the surrounding terrain, many regions might receive a high level of solar radiation throughout the year. Ground measurements confirm that most locations in India are receiving daily solar radiation levels between 4,000 and 7,000 watts, which is in line with a statewide satellite-based estimate. The sun's horizontal irradiation is strongest in the states of Rajasthan and Gujarat. Methods used in solar power plants may be broken down into many broad groups including a) Solar Photovoltaic (SPV) plants and b) Solar Thermal Power plants.

Additional types of solar photovoltaic technology include thin film solar cells, mono-crystalline Si solar cells, and multi-crystalline Si solar cells. To this day, crystalline Si solar cells make up between 85% and 90% of the market, with thin film technologies making up the remaining 10-15%. Although there are benefits and drawbacks to every solar cell technology, photovoltaics (PV) are now the sole option for reducing carbon emissions from the power system. Given that India is the world's third-largest emitter of carbon dioxide, this topic is of paramount importance in the Indian energy sector (after the USA and China). Research shows that solar photovoltaics have a far lower lifetime carbon dioxide emission rate compared to coal-fired power facilities. Solar power plants create the maximum CO_2

equivalent emissions during the production process, which explains this phenomenon.

Due to factors including high levels of solar irradiation, the need for rural electrification, the many benefits of solar power, and the expansion of the solar energy market, progressive policies have been established throughout the years to facilitate the expansion of the solar power sector. The numerous difficulties in creating regulations and implementing technology are discussed later in this chapter as well as the several different ways in which actions may be performed right now to improve the situation.

6.4 Solar Photovoltaic Market in India

India is well on its way to becoming a solar energy centre because of the country's immense potential for producing electricity from solar power. India is a prime location to take advantage of the technological and economic promise of photovoltaics. Assuming annual GDP growth of over 8%, the "gap" between energy supply and demand will continue to widen. Solar photovoltaics (PV) is a renewable energy source with the potential to close this gap. In most of the country, there are between 300 and 330 bright days annually, which amounts to more than 5,000 trillion kWh, or more than quadruple India's annual energy consumption. Typically, the sun hits a surface with an intensity of between 4 and 7 kWh/m_2 on a given day. Solar lanterns, home/street lighting systems, solar water pumps, etc., make up 80% of the one million industrial PV systems that have been deployed, with an installed capacity of roughly 66 MW. New and effective renewable energy technologies are being developed because of the Paris Agreement to minimize GHG emissions. The initial solar cells were costly and supplied electricity with relatively modest performance, but coal-based power was cheap and readily available. However, numerous studies have demonstrated the exponentially declining cost of solar technology. One study suggests that solar panel prices have reduced by as much as 80%. since just 2009 alone. The falling price of solar panels and the imperative to decarbonize the power grid have led to a dramatic increase in their utilization. There was a reduction in the price of solar photovoltaic systems from 2010 to 2015, from 0.26 USD/kWh to 0.285 USD/kWh. The International Energy Agency prepared a study in 2018 that details how inexpensive photovoltaic bids are in various regions. In Chile and Mexico, Levelized power costs achieved a historic low of 2.1 USD/kWh, while in Europe they decreased to 5 USD/kWh.

6.5 Need for Solar Energy

- **Energy Security**

  a) Nonrenewable resources provide most of India's energy needs.
  b) Because of their scarcity, renewable energy sources are becoming important.
  c) Solar power in India has the potential to meet the country's growing need for eco-friendly electricity.
  d) India must rely on imports to meet its energy needs, resulting in massive costs and unpredictability in this area.

- **Economic Development**

  a) India is a rising nation that requires reliable power to support its expanding manufacturing and agricultural sectors.
  b) India's economy and industrial sector will benefit from reliable, cheap electricity that is generated on-site.

6.6 Government Initiatives

a) There is a centralised government office in India responsible for coordinating all issues related to alternative and renewable energy sources.
b) The National Solar Mission is an ambitious program that was started by the national and state governments of India to promote sustainable development and satisfy the critical energy demands of the country as a whole.
c) The Indian Renewable Energy Development Agency (IREDA) is a non-banking financial agency that provides term loans for projects related to renewable energy and energy efficiency. It is the responsibility of this ministry to monitor the activities of IREDA.
d) The National Institute of Solar Energy is an independent organization that was founded by the Ministry of New and

Renewable Energy (MoNRE) to carry out research and development.
e) Improvements to the grid can be made in conjunction with the development of renewable energy.
f) Encourage the use of solar panels to generate electricity along canals and in canal storage facilities.
g) The Solar Rooftop Integrating Photovoltaics for a Sustainable India (SRISTI) initiative aims to facilitate solar energy installations on residential and commercial rooftops throughout India.
h) Programs like Suryamitra help get people ready for the job.
i) Large energy users are subject to a renewable procurement requirement.
j) Government initiatives like the Green Energy Corridor and the National Green Energy Program advance the use of renewable energy.

6.7 Challenges for Solar Thermal Systems

A major hurdle for solar thermal technology is harnessing the sun's rays to produce useful, storable thermal energy. By focusing sunlight into thermal power reactors, temperatures above 3,000 °C may be achieved, facilitating the effective chemical synthesis of fuels from raw materials. For solar thermal reactors to be practical, new materials will need to be created that can survive the extremely high temperatures that will be generated. Similar to how nuclear fission reactor heat may split water to generate H_2, concentrated solar thermal energy may be converted into chemical fuel using novel chemical conversion processes with amazing efficiency and cost-effectiveness. Lower sun concentration temperatures allow for the utilization of solar heat to power turbines that mechanically create electricity, which is more efficient than the existing solar photovoltaic production. It is possible that a solar-powered engine coupled with solar-powered chemical storage and release cycles, such as those based on the dissociation and synthesis of ammonia, could enable the creation of useable energy in a continuous manner around the clock. High heat storage capacities and longer release durations, spanning the diurnal cycle, may be possible using novel thermal storage materials that incorporate a phase transition. Using nanostructured thermoelectric materials such as nanowires or quantum dot arrays, it is theoretically possible to directly generate energy from temperature differentials with an efficiency of 20-30% across a temperature differential of a few hundred degrees Celsius. This direct energy generation could be accomplished using quantum dot arrays or nanowires. Since the differences are so much larger now, solar thermal reactors can function at considerably greater efficiency. Focusing systems for concentrated solar thermal technologies requires innovative high-performance, low-cost reflecting materials to reach their full economic potential.

6.8 Benefits of Solar PV

a) The development of renewable energy sources has been prioritized by many national governments for a variety of reasons, including the reduction of emissions and the achievement of international climate targets, as well as for broader socioeconomic advantages.

b) IRENA's modelling work that evaluates the socioeconomic consequences of the Remap scenario suggests that the solar sector will account for more than 11.7 million jobs by the year 2030 and more than 18.7 million jobs by the year 2050. This prediction is based on the organization's evaluation of the socioeconomic consequences of the Remap scenario. Decentralized off-grid solar PV is becoming increasingly popular, which is good news in places like developing nations and regions with spotty grid infrastructure. GOGLA and Vivid Economics (2018) made a recent estimate that these applications have created 372,000 new jobs in South Asia and other regions of sub-Saharan Africa. Employment may grow substantially in the coming years as a result of ongoing price cuts and strict implementation of equipment quality requirements.

c) Technical, business, administrative, economic, and legal expertise are just a few of the many that are in high demand as the solar energy sector continues to expand. Improving gender equality and justice aren't the only reasons to encourage more women to enter the renewable energy sector; expanding access to a larger pool of qualified candidates is another important aspect.

d) In 2018, the solar sector was predicted to employ 3.6 million people throughout the world. Employment in this sector has increased at a faster rate than any other renewable energy technology. In terms of solar photovoltaic (PV) jobs, about 3 million (or 85% of the global total) may be found in Asia, followed by 6.4% in North America, 3.9% in Africa, and 3.2% in Europe.

6.9 Causes of Delay in Solar PV Implementation and Ways to Quicken the Rate of Installation

This study reaffirms the importance of solar photovoltaics in bringing about the global energy transition required to achieve the Paris climate goals. This technology already exists, can be broadly implemented quickly, and costs less than competing options. Despite these encouraging signs, solar PV power projects still face serious obstacles that could impede the fast expansion needed over the coming decades. Variations in project requirements, local conditions, and technological development all contribute to uneven impacts on renewable energy. Current restrictions at multiple levels may slow the deployment of solar PV capacity during the next three decades (technological, economic, governmental, regulatory, and socio-political). It is crucial to address these issues right away through a range of legislative interventions and implementation tactics to ease future deployment. Although there may be positive socioeconomic effects from energy transformation on a global scale, a more in-depth examination of regional differences illustrates how the transformation's advantages (and disadvantages) might be felt in various regions of the world. The main points are as follows: 1) Countries begin their energy transitions in different places, 2) have supply chains that vary in length, strength, and diversity, 3) rely on fossil fuels to varying degrees, and 4) have varying degrees of national ambition and means of implementation.

A policy framework that works on two fronts is required to equalise the regional and national effects of the energy transition. The goal of lowering energy consumption and expanding access to electricity should both be supported by the framework. At the same time, energy policy as a whole needs to be rethought to make energy a driver of inclusive, long-term economic growth. Deployment policies, integration policies, and enabling policies are the pillars on which a "just transition" policy framework is built. To guarantee a fair and inclusive energy transition, these sets of policies would need to collaborate, primarily to remove all the obstacles that currently stand in the way (economic, technical, political, and social policies and regulations). There should be careful consideration of the circumstances and goals of each country when deciding on the instrument to use and how it should be constructed. Future measures and solutions are more comprehensive, necessitating a study dedicated solely to overcoming industry hurdles.

6.10 Future Trends of Solar PV

This section is intended to give insight into the technologies that are driving the solar PV industry, its continuous expansion, and its potential to significantly transform the energy system. As the market expands and diversifies, it faces new obstacles, which are investigated. Innovations are happening down the value chain in the solar PV business, which means change is quick. Greater efficiency has been a driving force for innovation in recent years. Because of this expansion, Technology like passivated emitter and rear cell/contact (PERC) is developing, which is a sign of this trend. To improve the efficiency of solar panels, it delivers more efficient solar cells. As the number of cells needed to process a given amount of output increases with module size, increasing cell efficiency is essential for cost-effective module production. At the system level, efficiency is also crucial, with several motivating factors supporting the recent surge in the development of more efficient technologies. Increased technological efficiency cuts down on the number of modules that need to be transported to the installation site, the quantity of land that needs to be utilized, and the length of wires and cables that need to be utilized. From a promotional standpoint, consumers will associate a higher degree of technical skill with a business that can provide the most efficient modules.

6.11 Conclusion

The circumstances in India are ideal for producing solar power. Millions of Indians, particularly the country's poorest residents, stand to benefit from the fast spread of solar electricity. It has the potential to undergird growth in all areas of development, as well as generate thousands of jobs in the solar sector, bringing the country closer to its goal of being the "India of the future." Rooftop solar panels remain a viable area for the use of solar energy via programs and initiatives that may help India break new ground in solar power generation and position the country to become a global leader in green energy. India's tropical location, large market, rich legislative incentives, and many educational and research institutes make it a viable candidate for global energy dominance. Insufficient equipment upkeep and difficult-to-negotiate service charges make government-provided energy expensive and unreliable. India's National Solar Mission aims to improve renewable energy access. The country's incapacity to maintain solar power plants may restrict this program's viability.

References

1. F. Jiang, H. Xie, and O. Ellen, "Hybrid energy system with optimized storage for improvement of sustainability in a small town," *Sustain.*, vol. 10, no. 6, pp. 1–16, 2018, doi: 10.3390/su10062034.
2. D. Kumar, "Economic Assessment of Photovoltaic Energy Production Prospects in India," *Procedia Earth Planet. Sci.*, vol. 11, pp. 425–436, 2015, doi: 10.1016/j.proeps.2015.06.042.
3. S. Chakraborty, S. Das, and M. Negi, *Hybrid Microgrids for Diesel Consumption Reduction in Remote Military Bases of India*, vol. 580, 2020.
4. A. Iqbal and T. Iqbal, "Design and Analysis of a Stand-Alone PV System for a Rural House in Pakistan," *Int. J. Photoenergy*, vol. 2019, p. 8, 2019.
5. D. Cannizzaro, A. Aliberti, L. Bottaccioli, E. Macii, A. Acquaviva, and E. Patti, "Solar radiation forecasting based on convolutional neural network and ensemble learning," *Expert Syst. Appl.*, vol. 181, 2021, doi: 10.1016/j.eswa.2021.115167.
6. Z. Hu, "When energy justice encounters authoritarian environmentalism: The case of clean heating energy transitions in rural China," *Energy Res. Soc. Sci.*, vol. 70, no. April, p. 101771, 2020, doi: 10.1016/j.erss.2020.101771.
7. A. Armin Razmjoo, A. Sumper, and A. Davarpanah, "Development of sustainable energy indexes by the utilization of new indicators: A comparative study," *Energy Reports*, vol. 5, pp. 375–383, 2019, doi: 10.1016/j.egyr.2019.03.006.
8. M. Sánchez, W. S. Ochoa M, E. Toledo, and J. Ordóñez, "The relevance of Index of Sustainable Economic Wellbeing. Case study of Ecuador," *Environ. Sustain. Indic.*, vol. 6, no. May, p. 100037, 2020, doi: 10.1016/j.indic.2020.100037.
9. L. Gao, X. Wen, Y. Guo, T. Gao, Y. Wang, and L. Shen, "Spatiotemporal Variability of Carbon Flux from Different Land Use and Land Cover Changes: A Case Study in Hubei Province, China," *Energies*, vol. 7, no. 4, pp. 2298–2316, Apr. 2014, doi: 10.3390/en7042298.
10. A. W. H. Poi, T. H. Law, H. Hamid, and F. Mohd Jakarni, "Motorcycle to car ownership: The role of road mobility, accessibility and income inequality," *Transp. Res. Part D Transp. Environ.*, vol. 90, 2021, doi: 10.1016/j.trd.2020.102650.
11. M. Santoleri, S. Müller, and M. Buchecker, "Head waters as essential green-blue infrastructure for local recreation? Analysis of general public usage

habits in aargau wiggertal [Gewässer als zentrale grün-blaue infrastruktur der naherholung? Analyse von nutzungsgewohnheiten der bevölkerung im aarga," *Naturschutz und Landschaftsplan.*, vol. 53, no. 8, pp. 22–29, 2021, doi: 10.1399/NuL.2021.08.02.

Other Works Consulted

1. IEA 2015, India Energy Outlook, US Energy Information.
2. Shahsavari A., Akbari M. (2018), Potential of solar energy in developing countries for reducing energy-related emissions, *Renew. Sustain. Energy Rev.* 90, 275–291.
3. Teske S., Sawyer S., Schäfer O., Pregger T., Simon S., Naegler T., Schmid S., Özdemir E.D., Pagenkopf J., Kleiner F., Rutovitz J. (2015), Energy [r] Evolution-A Sustainable World Energy Outlook .
4. Hairat M.K., Ghosh S. (2017), 100 GW solar power in India by 2022–A critical review, *Renew. Sustain. Energy Rev.* 73 1041–1050.
5. Moallemi E.A., Aye L., Webb J.M., de Haan F.J. (2017), George B.A., India's on-grid solar power development: historical transitions, present status and future driving forces, *Renew. Sustain. Energy Rev.* 69 239–247.
6. https://m.economictimes.com/industry/renewables/is-transitioning-to-solar-energy-for-electricity-in-urban-india-sustainable/articleshow/86678428.cms
7. https://MNRE.gov.in>Scheme-documents
8. https://www.researchgate.net
9. Ahmad F., Alam M.S. (2018), Economic and ecological aspects for micro grids deployment in India, *Sustainable Cities and Society* 37 407–419.
10. Saxena N. (2018), solar energy as renewable energy systems: perspective and challenges in Indian context, *Int. J. Eng. Technol. Sci. Res.* 5 (1).
11. Garg P. (2012), Energy scenario and vision 2020 in India, *Sustain J.. Energy. Environ.* 3 (1) 7–17.
12. International Renewable Energy Agency (IRENA), Quarterly Report: Letting in the Light (2016).
13. International Energy Agency (IEA) (2018), Report IEA-PVPS: Trends in Photovoltaic Applications.
14. Braun,S.,Hahn,G.,Nissler,R.,Pönisch,C.,Habermann,D. (2013), "The Multi-busbar Design: An Overview", *Energy Procedia* 43, 86-92, https://doi.org/10.1016/j.egypro.2013.11.092
15. Subhes C. Bhattacharyya, "Energy access problem of the poor in India: Is rural electrification a remedy?" *Energy Policy* 34 (18):3387-3397, December 2006.

7

Energy Management Strategies of a Microgrid: Review, Challenges, Opportunities, Future Scope

Chiranjit Biswas, Somudeep Bhattacharjee, Uttara Das and Champa Nandi*

Department of Electrical Engineering, Tripura University, Agartala, Tripura, India

Abstract

The generation of clean energy is crucial to sustainable growth and worldwide environmental concerns caused by climate change. Renewable energy sources have previously demonstrated their low or zero carbon emission generation efficiency. Now, the idea of a microgrid is to have a self-maintained system that consists of numerous energy resources, which can help to operate the microgrid during the collapse of the grid. As a microgrid utilizes numerous energy sources, the energy must be managed in a safe, smart, coordinated, and reliable manner. We have attempted to analyze some research papers to learn about the limitations of microgrid energy management systems and discover how to manage energy in a microgrid in a much smarter way. This document provides essential background for our ongoing efforts, as it details methods for controlling the power in microgrids.

Keywords: Renewable resources, energy management system, microgrid

7.1 Introduction

Today, managing energy is more important for us because generating and distributing energy affects our environment and our daily life. Generation through non-renewable sources like fossil fuels, coals, and various types of oil emits greenhouse gasses (CO_2 emission) that directly affect the environment by heating the atmosphere [1]. Sometimes generating units situated

---

Corresponding author: chmpnandi@gmail.com

far from the distributed system create power losses. Therefore, demanded power is not generated from the generating units. According to research, one day non-renewable sources will end due to the extreme number of non-renewable sources that are still used by generating units. So, it is necessary to manage the energy [2]. Energy management is the specific way which helps to fulfil our power demand, and it also gives an idea about the eco-friendly characteristics of our environment. In recent days, renewable sources like wind, solar, hydro, tidal, geothermal, and biomass energy are used for energy generation; those are more helpful for our environment and fulfil our energy demands easily. Renewable sources save the consumption of the cost related to raw materials and change the pattern of a clean cost structure. Renewable energy sources are also helpful for reducing the use of fossil fuels, coal, and oils in generation units. If we manage the power of renewable energy sources properly, then we get a more efficient power supply and we can also save the consumption of cost, input power, etc. Managing the energy in microgrids is more reliable than using renewable sources [3, 4]. Microgrids integrate the power supply of renewable and sometimes conventional sources. Energy management is used for reducing and controlling energy consumption. There are many techniques used for energy management of microgrids like AI (Artificial Intelligence) based, fuzzy logic–based, etc. By using those techniques, it is easy to manage the energy of the microgrids: the battery storage and carbon emission. Using those techniques, we can get eco-friendly, low-cost energy management systems in microgrids. These techniques have to control the power supply, simulate various values of generated energy, forecast values of generated energy and give proper energy management systems [5]. All those techniques are reliable for the energy management of the microgrids.

In 2018, Rafique *et al.* discussed the benefits of energy management strategies in a microgrid. They presented various control techniques for solving microgrid problems and discussed their advantages and disadvantages [6]. Zia *et al.* presented various decision-making strategies with their solutions for energy management on microgrids. The goal of that paper was to do overall sustainable development by optimizing the operation, scheduling the energy and making it reliable [7]. In 2019, Vera *et al.* focused on two different approaches based on optimization for managing the energy in the microgrid. Both approaches were used for the cost-benefit balance of microgrids [8].

In this study, we dive deep into the topic of microgrid energy management solutions, with an emphasis on the use of hybrid renewables. We also conduct a comprehensive literature review of the various hybrid systems based on renewable resources, detailing their advantages and disadvantages.

We also described the future work of different hybrid systems employed in a microgrid. Thus, this study will help to comprehend individual efforts for improving the future of energy management systems in terms of renewable energy by reducing carbon emissions and cost reduction. The second half of this study presents the research approach taken. This chapter highlights numerous energy management tactics utilized in microgrids with their full explanation in section 7.3. Issues with microgrid energy management are covered in section 7.4. Section 7.5 highlights the prospects for the usage of energy management strategies in a microgrid. The future research directions of energy management systems in microgrids are presented in section 7.6, and the future of the field is discussed in section 7.7. Figure 7.1 displays the classification system used in this article.

7.2 Methodology

In this paper, an in-depth study of microgrids, energy management systems, fuzzy logic algorithms and artificial intelligence (AI) is achieved.

7.2.1 Research Studies Selection Criteria

We have computed using the title, abstract, and keywords fields according to the database search. We have used Scopus, IEEE Xplore, and Google Scholar for searching relevant research. We have used energy management, microgrid, fuzzy logic algorithm and artificial intelligence as the search string. We have used research publications for the period 2014 to 2020.

7.2.2 Section of Literature

The systematic literature review shown in Figure 7.2 includes all criteria of the application. Firstly, we have taken 100 possible papers. Then we screened the titles and abstracts of relevant studies and works of those papers. After all the testing criteria, we chose 51 papers.

7.2.3 Testing Criteria

Testing criteria gave us the intent of the study's significance according to the relevant questions. The research study described the quality criteria and was selected for the systematic literature review.

01	02	03
INTRODUCTION	**METHODOLOGY**	**PRELIMINARY**
Preface	Research studies selection Criteria	Fuzzy Logic-based Management strategies
Some previous Literature Review	Section of literature	AI-based management strategies
Structure of Paper	Testing criteria	Others management strategies
	Extraction of data	
	Findings	

04	05	06
CHALLENGES	**OPPORTUNITIES**	**FUTURE RESEARCH DIRCTION**
Unpredictable atmospheric condition	Problem regarding cost	Low cost
Variation of load demmand	Problem regarding CO_2 emission in microgrid	Low carbon emission
Proper SOC control	Problem regarding uncertainty	Battery's SOC
Proper cost balancing	Problem regarding various storage system	
Proper balancing between grid and generating unit		

07	08
CONCLUSION	**REFERENCES**

Figure 7.1 Taxonomy of the paper.

7.2.4 Extraction of Data

Here we have categorized the chosen studies into two algorithms, a fuzzy logic algorithm and an Artificial Intelligence algorithm. This research data is more relevant to our research queries.

7.2.5 Findings

The taxonomy is used for synthesizing the research and categorizing the literature. The study is briefly described in each part.

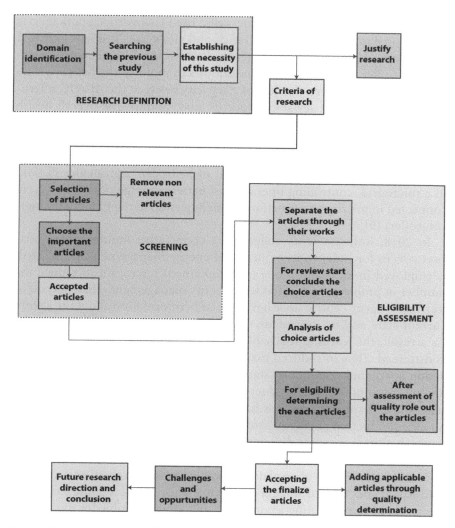

Figure 7.2 Literature review diagram.

7.3 Preliminary

7.3.1 Fuzzy Logic–Based Management Strategies

In 2019, Al-Sakkaf *et al.* displayed the plan for a DC microgrid, the functioning of which is determined by real-world information and lower-complexity inputs via an optimizer called Fuzzy Logic Energy Management System (FLEMS). In this research, their grid-off microgrid consists of photovoltaic, wind turbines, and fuel cells. The FC's hydrogen tank level is proportional to the SOC of the batteries, and vice versa. This study created a fuzzy logic–based energy management system. An FLC-reliant supervisory controller was proposed by the authors to monitor the battery's state of charge (SOC), govern the flow of electricity between the AC/DC microgrid and the active renewable resources, and monitor the overall health of the system. The Artificial Bee Colony (ABC) idea was utilized to optimize FLEMS for use with microgrids. In this paper, there are many limitations like improper economic dispatch generation and power flow to the microgrid, maximising the production of hydrogen, optimizing FLEMS in a microgrid, controlling operation of battery storage of those that are connected in microgrids, functions of membership in the FLC (fuzzy logic controller) [9].

In 2018, Kofinasa *et al.* suggested a cooperative multiagent foundation system for independent microgrid energy management. Here, MAS is employed in an island-mode (self-contained) microgrid to reduce the number of problems associated with energy management. This method of learning is agent-centric since it makes use of information about the agent's current state and its immediate environment to maximize performance. As a result, the learning mechanism's enhancement and state space were constrained. They introduced fuzzy-Q learning for every agent because of action space and continuous state. The balancing of the power problem among consumption and production units is solved by a decentralized system. The experimental result is based on concern demand load, real data, and the production of energy using photovoltaic sources. It is found in the paper that there are some difficulties in balancing the power between consumption and production units [10].

In 2018, Kofinas *et al.* explained a system of single-agent solar microgrid which goes toward solving energy management. It is going to resolve the issues of incorporating a consumer unit which is liable for giving service and accessories at an appointed level of demand. This gives a trade-off beyond good specification for the consumption and generation of energy. The authors presented the function of fuzzy reward which enhanced the mechanism of learning. They used the technique of reinforcement learning (RL) for providing quality demand service. Incorporated consumer units (those are answerable for providing goods and services at specific demand levels), and complexity in controlling the flow of energy between the solar microgrid elements are the problems of the microgrid. Here the good performance of agents does not come directly, it depends on learning and error technique [11].

In 2018, Ganesan *et al.* proposed a method to control and integrate the battery storage and PV source in the microgrid system. The goal of that paper was to minimize the distortion of the entire harmonic from the output. Using fuzzy logic and proportional-integral controller, they achieved the inverter control of the closed-loop. The authors presented the BESS (battery energy storage system) by including a DC-DC buck-boost bidirectional converter for many types of hold state, discharging and charging states. The limitations of this paper are complexities in controlling of close loop, charge and discharge state of batteries, harmonic distortion, etc. [12].

In 2018, Jafari *et al.* proposed a residential grid-tied smart microgrid that integrates PV energy using combined magnetic and electrical buses. Here, the local loads are supplied from a battery bank and fuel cells. In this paper, to reduce the current ripple high frequency, the author represented the interleaved method. This method will give results on the stability of the MPPT process. TAB converter ports that reduce RMS and peak current employ a dc synchronized bus voltage (SDVB). Feed-forward compensation of the inverter control loop is used to reduce the low-frequency ripple propagated from the inverter output voltage dc bus. Here, the article falls short due to the difficulty of stabilizing the MPPT while also managing the battery's power flow [13].

In 2018, Al Badwawi *et al.* explained how the energy and battery power do not go beyond the limit of design and maintain stability in power flow. In this research, EMS (Energy Management System) was developed for islanded AC microgrid. The microgrid's frequency is used to create a wireless version of the supervisory controller. The system performance need is met by taking into account the design, the maximum power of the battery restrictions, and the SOC. The paper's authors set out to reduce the run time of the unit's auxiliary power supply while still meeting load requirements and keeping battery costs low. In this study, we simulate real-time operations in a variety of settings, verify the controller through experiment, and contrast it to a conventional controller. Here, the limitation of the paper was controlling the battery discharging/charging power and SOC (state of charge) is not always possible to maintain by the proportional controller [14].

In 2018, Jafari *et al.* proposed a residential application of a microgrid that is operated by a novel EMS with two horizons. The study analyzed the effectiveness of a real-time fuzzy controller used in energy management units (EMU) of two different sizes. In this study, we focus on four important parameters: the forecast profile and real-time price of grid electricity; the demand for load power; the generation of PV power; and the total amount of power used. The authors also highlight the power flow performance of efficiency of the converter and battery; the running and capital energy cost of fuel cells and PV cells. They provided a detailed study of microgrid control and energy management operation modes with a large number which include control of real time, the strategy of mode selection, and using the function of hysteresis, which is based on fuzzy membership oscillation of mitigating transient and defining modes of bridging. From the paper, we find the current rating and electrical power are the operating limits, fluctuation of grid power is complex to control and, due to the large operation mode numbers it is complex to manage the energy distribution and manage the cost [15].

In 2018, Prathyush *et al.* suggested the idea of limiting power variation while keeping battery charge conditions within the limit of tolerance by enhancing power profile fluctuation and decreasing FLC complexity. In this work, we use renewable energy sources like WT and PV panels. To progress the performance of the system and stability, they used elements of energy storage like (an EMS) energy management system and elements of storage (ultracapacitors, batteries, etc.). The authors used LPF (low pass filter) to isolate the elements of the frequency which is low and frequency which is high of the microgrid. They also showed the EMS system of FLC using 25 rules, one output and two inputs. The design allows setting an 80% battery SOC of rated capacity. From this paper, it is found that the power fluctuation controlling in the grid is not done; it is complex to reduce the microgrid operating cost and due to the fast reaction to energy changes of the microgrid, it is complicated to set the SOC of the battery at rated capacity [16].

In 2018 Ji *et al.* introduced a strategy for the management of energy in residential CCHP (combined cooling, heat, and power) microgrid. For this, the thought of the demand for uncertain energy, the output of wind power, the penalty of CO_2 emission and electricity price are used here. The daily cost of the approach and the emission of CO_2 are both reduced with the use of the CCHP microgrid system in this study. An explicit interval of the fuzzy risk model of linear programming is applied here for the planning of recycling of vehicle's end-of-life in the EU. Proposed methods have more advantages, such as 1) it is not one heir the advantage of specification programming of the interval, but also gleams the clear trade-off data of explicit cost-risk. 2) The equivocal attitude of risk since decision-maker occur to allow for, which makes the decision procedure more realistic. Here, the limitation was on complex controlling of the CO_2 emissions, electricity price, and the demand for uncertain energy [17].

In 2017, Arcos-Aviles *et al.* proposed a one-grid residential microgrid which is constructed based on Renewable Energy as resources and Storage systems as a battery (BESS). The author aimed to smooth the operation of the microgrid here. In a fully isolated microgrid, the primary goal of the management system is to maintain constant client power. Controlling the ability outcomes once required and generally victimization Demand Side Management (DSM) methods to ignore depletion of battery are the most important points for EMS. For on-grid conditions, wherever this assignment concentrates on the grid, which may work as the power supply or an influence sink, make sure the feasible power is provided to shoppers. For this event, the EMS must capture the ability of flow on the microgrid parts to achieve a collection of predefined objectives like minimizing the microgrid operational prices or increasing the revenues in line with weight unit bids and electricity value. Here we get that it is not suitable for grid power control due to power forecasting problems and the State-of-Charge (SOC) of batteries [18].

In 2017, Angalaeswari *et al.* proposed a competent energy management framework that uses the Fuzzy Logic Controller (FLC) to generate the duty ratio of the maximum power point tracking controller. In this study, a grid-connected battery and solar array are employed with a fuzzy logic maximum power point tracking (MPPT) controller. FLC conveniently handles the current and voltage level of the grid, allowing the irradiation to be changed for result validation. This paper reveals that the MPPT is tricky to regulate [19].

In 2016, Yang *et al.* introduced a microgrid energy management technique to minimize the expenditure of PV and BESS. In this paper, PV modules were the predominant contributor to the expenditure of the system. Here, a seaward propensity was added to the improvement of silicon technology including Thin Film technologies. Growing emulation between the manufacturers and a large enlargement within the generation capability of PV modules were the main topic of discussion in this paper. From this paper, we find that the energy storage system is not easy to control, and the wind generator is not suitable for this strategy [20].

In 2016, Moafi *et al.* evaluated the execution of microgrids in on-grid and off-grid modes. In this paper, the proper overall working of the mentioned controller was to enhance microgrid balance and to connect the error that befell the EMS device. The design of the controller was based on a traditional controller and fuzzy-based PID. The superior stabilization of the mentioned controller was below numerous disturbances-specific implemented scenarios in the system. Adjustments in renewable resources result in an improved dynamic reaction of the device to minimize the deviation and oscillation of frequency. Microgrid frequency control is used to enhance transient balance by retaining the symmetry between generation and consumption in the strength management system. Here we find some difficulties like error creating in this process and process isolating, and load switching at the time of particular process execution [21].

In 2012, Chen et al. designed and implemented an EMS in a DC microgrid which is controlled by fuzzy logic. MATLAB/Simulink was used for modelling the proposed system here; it also helped in analysing and controlling distributed energy resources and storage systems. The completed monitoring energy management system (EMS) was performed by LabVIEW software. In this work, the authors optimize power flow using the developed controller and establish the SOC of energy storage per detailed requirements. A distributed intelligent energy management system for microgrids is simulated in hardware. An enhanced HILS testbed for a microgrid-based distributed intelligent control system using the Multi-Agent System (DIMS). Here we find that renewable resource does not give sufficient energy to the load. Thus, the storage system must work in a switching manner [22].

In 2012, Kyriakarakos et al. developed a Fuzzy Logic Energy Management System (FLEMS) for a microgrid with multiple independent power sources and tested it through simulation. In this paper, an accumulation of understanding from professionals and facts to design the fuzzy instruction base of the fuzzy logic framework is used. The FLEMS is gradable and supports plainness, and it is philologically explainable. A short statement of the models of Fuel Cell, Electrolyser, and Desalination devices was given from experimental check outcomes. The evaluation of the ON/OFF method and FLEMS technique is also included here. We find that it does not properly support the partition of load execution on the system [23].

In 2012, Chaouachi et al. introduced an EMS to reduce the Microgrid environmental effect and operation cost by taking its variables of preoperational load demand (LD) and renewable energies' future availability. In this paper, NNE (neural network ensemble) module is used for achieving high forecasting accuracy, improving the noise tolerance and generalization of machine learning. In a microgrid, the MIEM (Multiobjective intelligent energy management) can dispatch DG sources with MO for joint battery scheduling validation by comparing three approaches. An intelligent energy management framework is utilized to minimize microgrid emission levels and operating costs. To decrease the battery scheduling, a fuzzy expert framework is utilized. Here we demand that the Battery schedule planning of microgrid operation, forecasting accuracy, uncertainty of machine learning, emission level, and operating cost is complex to maintain [24].

7.3.2 AI-Based Management Strategies

In 2018, A. Tabanjat et al. studied a Hybrid Power system comprising Photovoltaic panels, Wind Turbines, Gas Micro-Turbine, and Fuel Cells associated with the energy storage of hydrogen. Here are discussed two (AI) artificial intelligence-based solutions which are used for the operational performance of energy management during load variation of 24h. Neural Networks (NN) and Fuzzy Logic (FL) are two energy management strategies. The suggested system was unable to capture all the demand meeting places due to the unpredictability of the battery, the demand of load, the irregularity of Renewable Energy Sources, and the irregularity of the pricing of the sources [25].

In 2015, Diamantoulakis et al. dealt with excessive data size and smart grid big data management techniques. The dynamic energy management (DEM) phasing of this big data management was set up in a network of Smart Grids (SG). Finally, they discussed various methods and techniques for further exploring the substructure system of forecasting and monitoring the real time. Because of application design, big investment, and lack of strategy, the growth of the system of big data applications in real is very slow [26].

In 2019, Roy et al. evaluated DR (demand response) for evaluating the response of customers due to additional indices and RNN utilising. The proposed algorithm was able to get together the necessary demand of load with the minimum cost of energy. In this paper, the advancement issue of multi-target is changing into an easy sub-issue that can be outfitted and utilized as the assumption of a bilateral procedure [27].

In 2019, Ma et al. created a system for predicting microgrid electricity needs for the day by analyzing historical data on electricity consumption and weather forecasts. The proposed strategy uses a combination of Wavelet Transform (WT)–based SA networks and Feedforward WT networks (FFANN). Data from a Beijing, China, microgrid that measures electricity use confirms the validity of the viewpoint. The scientists noted that the high forecast error is related to high electricity use, and uncertainty throughout the summer and winter months [28].

In 2019, Ghorbani et al. offered a multi-agent strategy for managing energy in a microgrid that incorporates renewable energy sources like wind and solar. With the important microgrid cost in mind, this EMS was developed with reaction to demand in mind, taking into account uncertainty in load and generation based on different scenarios for the microgrid energy management such as lightning search algorithm. Because of the PV problem and at night a time no sunlight, the unstable renewable energy production is requiring substantial energy storage facilities. It is needed to overstep the facilities of current storage by far. But they don't use a system of storage [29].

In 2018, Chui *et al.* examined the load monitoring system of non-intrusive and smart metering to make a case study of the electricity consumption of the appliance. Due to these findings in that context, research on smart cities and energy sustainability in urban space was added. This research paper is best in debate contribution on the utilization of energy and urban space sustainability which integrates IoT, artificial intelligence, and big data analytics. But because of the unimproved accuracy of the classifier, all brands of electrical appliances are not covered due to the measurement [30].

In 2017, Ganesan *et al.* studied a solution for robust energy management that will make easier the economic and optimum control of flows of energy in a network of the microgrid. Through the generation forecasting of renewable energy, this study enables to define the power flows management, predicts the available energy in batteries, and supplicates the appropriate operation mode based on the demand of load to bring out the economic and efficient operation. The convolution lies in the load smooth transition allocation to a Battery Energy Storage System (BESS) from the PV source [31].

In 2018, Yan *et al.* studied collecting the data for analysis of system uncertainty, assessment of system uncertainty, and OR dispatching. The authors used Artificial Neural Networks (ANNs) for both forecastings of load and PV power, choosing the output and input of ANNs. During the obtaining of the input and trained ANNs, load demand and day-ahead PV power were forecasted. All the collected data used here are predictive. In this paper, the main articulation is the collection of data for analysis of the uncertainty of the system, assessment of System uncertainty, OR transmission. Here, extra energy stored batteries are discharged during the nighttime. Limitations were found in forecasting uncertainties of RES, urban microgrids, or calculation [32].

In 2018, Panbao *et al.* proposed a method of optimization for the HESS in a microgrid (DC) EMS (Energy Management System) with several objectives was developed (Hybrid Energy Storage System). The authors chose an MMDP (Maximum Membership Degree Principle) fuzzy control approach as the most effective means of finding a resolution. The best solution achieved is the reliability and economy of the microgrid. To improve the non-dominated of the multi-objective and DE algorithm, they used microgrid multiple pare to obtain the solution of operation. Islanded and grid-connected microgrid optimization modelling improves the DE algorithm, reliability, and economy [33].

In 2019, Alhussein *et al.* built a model of a Multi-Head Convolutional Neural Network (MH-CNN) to predict wind speed and sun irradiance shortly. Appropriate electricity generation from solar panels and wind turbines is approximated using weather predictions. The energy produced by wind turbines and solar panels was estimated by formulating the wind speed forecast issue and the solar irradiance prediction problem, respectively. For evaluating various models of machine learning performance, models of persistence and smart persistence, they selected the model of baseline. Limitations associated with this paper were the prediction problem of wind speed and solar irradiance, model effectiveness, etc. [34].

In 2017, Baghaee *et al.* presented a system for hybrid photovoltaic/wind generation. The system of storage of hydrogen energy including an electrolyser was described, along with an algorithm for multi-objective management of optimal power. Microgrids use a hydrogen storage tank and fuel cell to generate electricity on demand. The reliability and affordability of the microgrid system were both improved with the addition of PV and wind generation technologies. The electrolyser at ESS uses hydrogen to transform electrical energy into chemical form. Particle swarm optimization (PSO) is employed here for hybrid stand-alone system sizing. This paper covers the limitation on the units of a generation that are intrinsically non-dispatchable, wind turbine generators, failure of DC/AC converter, and photovoltaic arrays [35].

In 2019, Murugaperumal *et al.* proposed a technique of hybrid system of microgrid system and energy management modelling. For microgrid optimal energy management, the approach of ANFMDA is presented. ANFIS is for load demand utilization and MDA is for cost-minimizing approaches used here. Their proposed model can't take the high electrical load; it has determination issues because of various load demand values, and high annual total cost [36].

In 2011, Sanseverino *et al.* used the concept of minimizing carbon emissions and reducing cost production and quality maximisation to improve the sustainability of microgrids. Their proposed concept focused on solving the 24-hour energy generation dispatch by allowing the central controller to monitor the executed plan for the execution system. This will interrupt new information of input and at every time interval, it will repair the executed plans. In this paper, the carbon emission is consumed every 24 hours [37].

In 2019, Phan *et al.* developed HRES (hybrid renewable energy system) with hydrogen FCs with battery combination. The author presented the sizing of the energy system of hybrid renewable hydrogen storage by using HOMER software for BascoIsland, Philippines. It is clear to us that when variables increase in a system, then it is not easy to solve the difficulties in the case of AI techniques. Therefore, in the control system, mechanisms of communication between agents are needed to improve. But, it is difficult to control the strategies of HRES (hybrid renewable energy system) [38].

7.3.3 Other Management Strategies

In 2018, Indragandhi *et al.* explained the hybrid microgrids design which is serviceable for quality and system price assessment. They employed MOPSO (multi-objective particle swarm optimization) to examine the optimization of AC/DC microgrid management. In this work, they showed how the created (photovoltaic) energy might boost the efficiency of a microgrid in one of fifteen distinct ways. Here, a Fuzzy controller was used for determining the discharge and charge of the bank of the battery. Applied MOPSO methodology was used here to define the power loss and energy cost. From this paper, we find that the implementation cost structure of a microgrid is too complex and it is difficult to analyse the behaviour of the supplier and the consumer in remote areas [39].

In 2018, Utkarsh *et al.* explained the scheduling of smart microgrid internal devices and trading of energy development strategy. In this paper, for developing a strategy of macro scheduling by optimization of real time, microgrid scheduling internal device simultaneously used trading power. A topological constraint of underlying networks is fully subsumed here, which gives an attainable solution in practice. The system of the multi-microgrid distributed solution is proposed here. In this proposed system, each agent of a microgrid can meet up with each neighbouring surrogate and independently reach the solution, using an algorithm of distributed algorithm based on a CI-base. The efficacy of this algorithm is demonstrated by benchmarking it against a distributed algorithm of state-of-the-art. Apart from trading of reactive power, trading of active power among microgrids was also contemplated. This helps microgrids improve network voltage and reduce power losses. Finally, we find that the optimisation of real time, incorporating the underlying network topology constraint and developing network voltage, is too complex [40].

In 2016, Bhattacharya *et al.* presented a multistage stochastic programming model to obtain procurement and strategies of storage for grid-connected microgrids. Here, a novel model of multistage SP (stochastic programming) is used to obtain viable procurement and strategies for the storage operation of an on-grid microgrid. In this paper, the authors developed the SDDP (stochastic dual dynamic programming) algorithm to obtain a high-quality solution for a horizon planning of 24 hours. From this paper, we find that the prices of real-time electricity, renewable generation, and demand response are too complex to maintain. It was very difficult to reduce the total cost of microgrid here [41].

In 2017, Yan *et al.* established a linear model to evaluate the classic parts like fuel and capital costs and reliability costs including lifetime cost, etc. This paper aimed to investigate the design and operation of microgrid optimization using branch-and-cut to solve the problem of markovian. To evaluate the lifetime cost of microgrids, the authors established the linear model. They used an optimized design strategy to limit device size to possible combination numbers. The method of efficient optimization and appropriate model-delivered optimization gave results accurately. From this paper, we find that the method used for evaluating the lifetime cost of microgrid, carbon emissions, and the total cost of the system was too complex [42].

In 2018, Romero-Quote *et al.* explained a mathematical formula for EMS (energy management system) in an isolated microgrid. This formula addresses the uncertainty using the method of affine arithmetic (AA). In this paper, to solve the problem of AAUC (UC of AA-based), AA-base EMS and economic dispatch of real time were used to process the novel dispatch. To compare with the presented approach of microgrid for deterministic benchmark, stochastic, MPC, and techniques of stochastic-MPC, robust, and cost-effective solutions, the authors have provided the approach of AA-based EMS. From this paper, we find that the economic dispatch of real time, the overall system cost is too complex to control [43].

In 2018, Wang *et al.* explained the developing strategies of two-stage energy management under the presence of highly renewable resources. In this paper, the multi-microgrids developing strategy of hybrid energy management system improves the layered privacy and accommodates the capability of RESs fluctuation by enhancing the cause of each microgrid and it is regulated by a microgrid community (MGC) network. Here, the dispatching and scheduling were influenced greatly in the microgrid because the decision-maker is risk-based. To minimize the operation cost, a day-ahead scheduling control based on the method of hierarchical optimization is used. From this paper, we found that the total operation cost, risk of low profit, and many uncertainties in the dispatch stage of real time are too complex to handle [44].

In 2018, Thirugnanam *et al.* explained the energy management strategy of batteries used in the microgrid, where the DGs and PV are the major electricity source. In this paper, the authors designed a model for controlling the level of PV and DGs system. BEMS operation is used for operating DGs and battery characteristics simultaneously. The operation mode of BEMS is achieved by reducing the operating hours of DGs, reducing the power fluctuation of PV, extending the lifetime of the battery through battery discharge controlling and/or charge rate, and many batteries' coincident management of the different characteristics. It is done for verifying the BEMS using power generation of PV and real-world values data. From this paper, we find that to reduce fossil fuel of DGs consumption, and PV power fluctuation, the life cycle of charge and discharge of the battery is too complex to control [45].

In 2018, Netto *et al.* analysed the real-time management problems of the smart grid which is integrating the energy management of a power system by a telecommunication system. In this paper, for all agents' integration and for translating the information between MAS and the power system, IP/TCP communication system is used. For better analysis in real-time modelling, MAS uses two different triggers. For the polling mechanism problem, the authors add an event-driven mechanism. From this paper, we find that reducing greenhouse gases, carbon emissions, and real-time framework analysis is too complex to maintain. It is complicated to create a bridge for data communication between MAS and the power system [46].

In 2019, Jafari *et al.* presented an energy management topology of microgrid which is based on the current standard of the suburban system of renewable energy. In this paper, the author focused on increasing the operation and flexibility and decreasing the control complexity of the microgrid, isolating the converter port for the requirement of safety using a magnetic link associated with the component of the microgrid. The authors used the technique of synchronized bus-voltage balance (SBVB) for reducing the (Root mean square) RMS and a winding currents peak value of the magnetic link and device of switching which reduced conduction loss. The authors used a block of compensation in the loop of inverter control for reduced low-frequency present ripple generation since the O/P of the inverter and the bus of ac to the bus of dc and more distant to PV and buses of the fuel cell. This can submit high MPPT performance and filter components are a smaller size. A proposed microgrid can operate in a grid-connected mode and can operate in the operation mode of off-grid and compare different scenarios of energy management with the previously reported system. Novel energy management based on real-time data and prediction of long-term PV generation and demand of load using programming of 2D dynamic is proposed here. The microgrid component efficiency performance like converters, battery loss, and transformer is incorporated in the algorithm of optimization. The strategy of mode transition by a defined STD (state transition diagram) and for its smoothness, the transition of mode is employed as a bridging mode. From this paper, we find that the energy distribution and total cost of the system are high and cannot be limited to a preferable limit [47].

In 2018, Nemati *et al.* created a system that gives each DDG a respectable chance of being activated (DDG). The proposed method depends on the priority list method, which is tied to being online in a time step and a period. High-quality scheduling was established, with the primary goal being the reduction of total operating costs. These costs include things like fuel, renovation, network, and ageing expenses, as well as market transactions and fuel prices a day in advance. At the same time, it will work to cut down on the microgrid's greenhouse gas output. To overcome the limitations of the MILP algorithm, this work presents a new approach to managing topological restrictions inside communities. Here, numerous case studies are presented to highlight the inaccuracy of the optimization process and the dearth of sufficient results validations for optimization methods across a variety of optimization categories [48].

In 2018, Cardoso *et al.* suggested a minimal amount of human input is needed to define the area of possible solutions, which is guaranteed to be optimal. In this study, we account for the impacts of battery ageing to circumvent a shortcoming of multi-energy MILP models. The linearized formulation was initially introduced in this study to take battery ageing via IoT into account. Power natural philosophy-based converters and inverters facilitate the cost-effective economic conversion of inexperienced energy sources like solar photovoltaics and wind turbines. The paper's biggest unknown is the potential for switching devices to be damaged by fluctuations in frequency and voltage caused by unexpected power imbalances [49].

In 2017, Vergara *et al.* proposed a model for keeping an active power balance when PV generation exceeds predicted load consumption in the islanded mode of microgrid operation. In this study, they utilized a highly detailed MINLP model for the EM downside, one that factors in PV, ESSs, Gus, and cargo management. The microgrid is a model in association with nursing unbalanced, three-phase EDS, making the suggested model unique in comparison to previous efforts. The authors' failure to account for uncertainty in renewable generation and load consumption is a drawback [50].

In 2019, Tabar *et al.* investigated strategy for maximizing renewable energy generation while minimizing cost. The final advised harm by the buyer is thought to be acknowledged by the supplier due to the energy price for each amount, as stated in the comprehensive contracts between the client and provider. The varied behaviours of thermal and electrical masses, such as distinct peaks and value curves, are important in laying a solid foundation for the procedure. One problem with ES is that, as a result of shiftable load, not all of it is moved to the initial time intervals [51].

In 2013, Marzband *et al.* researched and investigated various energy management system scheduling operations and optimal operating strategies. Independent Wind Turbines (WT), Photovoltaic (PV) Cells, Microturbine (MT) Generators, and Energy Storage (ES) are all part of the system under scrutiny here. For microgrids operating in islanding mode, this work introduces a mixed-integer nonlinear programming (MINLP)-based EMS algorithm. A major breakdown is imminent because the system relies on electricity, which is threatened by the unexpected loss of micro resources and an accompanying spike in demand for that resource [52].

In 2015, Cheddadi *et al.* designed a LabVIEW application for a microgrid that can manage and supervise the whole system energy locally and remotely with efficiency. This system can control many mixed sources of energy like electrical phenomenon panels, turbines, diesel generators, and Battery pack equilibrates with efficiency, and it can assemble the loads (consumption). The major drawback of this application was that the particular designed system is not suitable for highly intermittent weather conditions and it was difficult to predict precisely by this system [53].

In 2020, Dabbaghjamanesh *et al.* used the reconfiguration approach to determine the best microgrid setup and avoid violating any constraints. The proposed method was created with the impact of the Dynamic Thermal Line Rating (DLR) taken into account. Using this paper's usual premise, an efficient dispatching solution may look significantly different from other methods. Fast data was guaranteed by the authors' use of cloud-fog computing on three levels. They devise a method of linearization to make the problem's nonlinear constraints linear. A major flaw in their suggested model is that DLR could be impacted by the line ampacity of the NMG, particularly in severe weather under the islanded mode, when the distribution line operates at its full capacity [54].

In 2017, Nandipati *et al.* developed a plan for utilizing IoT to facilitate communication between security and energy management systems. Power electronic switching device advancements help the power grid in this case by compensating for reactive power and enhancing power quality. Switching devices could be damaged by the abrupt changes in frequency and voltage caused by power imbalance [55].

In 2016, Chen *et al.* developed a model to fairly allocate the microgrid clusters' cost-cutting choice for maximum individual benefit. To investigate transitive energy management for microgrid clusters, a mathematical decision framework is provided here. This study proposes models that can be used to calculate the optimal local pricing strategy for energy. In the absence of a paradigm that strikes a balance between collective and private interests, the concept of microgrid clusters cannot be transformed [56].

In 2019, Mansour-lakouraj *et al.* developed a model to lessen the impact of natural disasters on operational costs and load reduction. To make the most of DERs, ESS, and DRs during both routine and emergencies, a novel risk-constrained day-ahead scheduling approach is introduced. We present a system with linearized AC power flow limitations to control voltages and energy flows within safe parameters. When severe weather strikes, the distribution system can be severely damaged, forcing the neighbouring microgrid operator to disconnect to meet demand locally [57].

In 2020, Shaterabadi *et al.* advocated the use of stochastic planning to make the offered strategy more robust and to more likely recover from the worst-case scenarios. Turbines made by INVELOX were used. In comparison to traditional wind turbines, INVELOX has various benefits, including a sixfold increase in power generation, the ability to operate at low speeds, decreased environmental impacts, lower investment costs, and less maintenance. Microgrid planning can take advantage of the turbine's benefits, and many scenarios involving or excluding INVELOX turbines can be considered. Due to a variety of limits and restrictions, the objective functions in this work are not optimized [58].

7.4 Challenges of Energy Management in Microgrids

1. We have seen that there are many problems like Electrical load, various load demand values, annual total cost, Co_2 emissions and electricity price, control of the output of wind power and control demand of uncertain energy.
2. Incorporated consumer units (those are answerable for provided goods and services at specific demand levels), complex to jurisdiction the flow of energy in the middle of the solar microgrid elements, the good performance of agent does not come directly, it depends on learning and error technique.
3. Somewhere we got the problem of controlling of close loop, charge and discharge state of batteries, harmonic distortion, and stabilizing the MPPT.
4. Battery discharging/charging power and SOC (state of charge) are unable to maintain by a proportional controller all the time.

5. In some of the cases like high consumption of electricity and uncertainty in the summer and winter seasons, the forecast error is high.
6. The problem of no irradiation in the nighttime and photovoltaics. The incalculable power generation into the bargain and the unstable/discontinuous production of renewable energy requires important facilities of storage for energy.
7. The current rating and electrical power are the operating limits of the system. Due to the large operation mode numbers, it is complex to manage the energy distribution and manage the cost.
8. Sometimes we got the power fluctuation controlled in the grid, reducing the microgrid operating cost; due to the fast reaction to energy changes of the microgrid, it is complicated to set the SOC of the battery at rated capacity.
9. The convolution lies in the load smooth transition allocation to a Battery Energy Storage System (BESS) from the PV source.

7.5 Opportunities

We can solve the problems like Electrical loads and various loads. Proper demand values, stabilize annual total cost, reduce CO^2 emissions, and normalize the electricity price. We can control the output of wind power by controlling the demand for uncertain energy. To govern the flow of energy between the components of the solar microgrid, we need to simplify the process of accounting for the goods and services given at certain demand levels. We were given the challenge of settling the close-loop control, battery charge/discharge states, harmonic distortion, and MPPT stabilization. The Battery's power and SOC (state of charge) must be managed by the proportional controller as it is discharged and charged. Due to low electricity use and seasonal fluctuations, the forecast error should be small. Limiting the system's operation to the maximum allowable current and electrical power ratings are recommended. Reducing the microgrid's operational cost and taming power outages are two priorities. The complexity arises when deciding how much of the PV-generated load should be transitioned smoothly to a battery-powered energy storage system (BESS). There are many opportunities we can implement in microgrids that possibly make the energy management system more reliable and more efficient.

7.6 Future Research Direction

We have seen that there are many problems like Electrical load, various load demand values, annual total cost, CO_2 emissions, electricity price, control of the output of wind power and control demand of uncertain energy. Incorporated consumer units, complex to jurisdiction the flow of energy in the middle of the solar microgrid elements, the good performance of an agent does not come directly, it depends on learning and error technique, somewhere we got the problem of controlling of close loop, charge and discharge state of batteries, harmonic distortion, stabilizing the MPPT. Battery discharging/charging power and SOC (state of charge) are unable to be maintained by a proportional controller all the time. In some cases, with high consumption of electricity and uncertainty in the summer and winter seasons, the forecast error is high. The current rating and electrical power are the operating limits of the system. Due to the large operation mode numbers, it is very complex to manage the energy distribution and manage the cost. Sometimes we got the power fluctuation controlled in the grid, reducing the microgrid operating cost, due to the fast reaction to energy changes of the microgrid, it is complicated to set the SOC of the battery at rated capacity. Furthermore, we have decided on a microgrid energy management system to solve the above problems relating to CO_2 emission, battery storage system, uncertain energies, distribution system, etc., so that it becomes more eco-friendly, reliable in continuous power supply and maintains the total cost of the management system.

7.7 Conclusion

Sustainable development and worldwide environmental concern due to climate change require clean energy generation. Renewable energy has proven effective for low/zero carbon emissions globally. The hybrid renewable sector faces challenges owing to weather, but they can be overcome using a microgrid. Energy management limits microgrid power peaks and shifts. Microgrids use demand and production predictions for energy management. Distributed generation (DG) technologies are crucial to today's micro/smart grids, and energy storage is a key component. This work summarises prominently published research papers that are much needed for our future work because we need to get knowledge about the energy management systems in microgrids by AI techniques, Fuzzy logic techniques, and some other techniques. We have also analysed different clean energy conversion technologies for energy management systems. We have studied some formulations of the scheduling models of some intelligence methods. We have analysed to investigate some different models.

References

1. Nandi C, Bhattacharjee S, Chakraborty S. Climate Change and Energy Dynamics with Solutions: A Case Study in Egypt, in: H. Qudrat-Ullah, A. Kayal (Eds.), *Climate Change and Energy Dynamics in the Middle East, Understanding Complex Systems*, Springer, 2019, pp. 225-257.
2. Bhattacharjee S, Das U, Chowdhury M, Nandi C. Role of Hybrid Energy System in Reducing Effects of Climate Change, in: H. Qudrat-Ullah, M. Asif (Eds.), *Dynamics of Energy, Environment and Economy, Lecture Notes in Energy 77*, Springer, 2020, pp. 115-138.
3. Olatomiwa L, Mekhilef S, Ismail MS, Moghavvemi M. Energy Management Strategies in hybrid renewable energy systems: A review. *Renewable and Sustainable Energy Reviews*. 2016;62:821-835.
4. Bhattacharjee S, Nandi C. Design of a voting based smart energy management system of the renewable energy based hybrid energy system for a small community. *Energy*. 2020;214:1-15.
5. Bhattacharjee S, Nandi C. Design of a Smart Energy Management Controller for Hybrid Energy System to Promote Clean Energy, in: A. Bhoi, K. Sherpa, A. Kalam, G.S. Chae (Eds.), *Advances in Greener Energy Technologies, Green Energy and Technology*, Springer, 2020, pp. 527-563.
6. Rafique SF, Jianhua Z. Energy management system, generation and demand predictors: a review. *IET Gener. Transm. Distrib*. 2018; 12(3):519-530.
7. Zia MF, Elbouchikhi E, Benbouzid M. Microgrids Energy Management Systems: a Critical Review on methods, solutions, and prospects. *Applied Energy*. 2018;222(C):1033-1055.
8. Vera YE.G, Dufo-López R, Bernal-Agustín JL. Energy Management in Microgrids with Renewable Energy Sources: A Literature Review. *Appl. Sci*. 2019; 9:3854.
9. Al-Sakkaf S, Kassas M, Khalid M, Abido MA. An energy management system for residential autonomous DC microgrid using optimized fuzzy logic controller considering economic dispatch. *Energies*. 2019;12(8):1457.
10. Kofinas P, Dounis AI, Vouros GA. Fuzzy Q-Learning for multi-agent decentralized energy management in microgrids. *Applied Energy*. 2018;219:53-67.
11. Kofinas P, Vouros G, Dounis AI. Energy management in solar microgrid via reinforcement learning using fuzzy reward. *Advances in Building Energy Research*. 2018;12(1):97-115.
12. Swaminathan Ganesan VR, Umashankar S, Sanjeevikumar P. Fuzzy-Based Microgrid Energy Management System Using Interleaved Boost Converter

and Three-Level NPC Inverter with Improved Grid Voltage Quality. *Advances in Smart Grid and Renewable Energy.* 2018:325.
13. Jafari M, Malekjamshidi Z, Lu DD, Zhu J. Development of a fuzzy-logic-based energy management system for a multiport multioperation mode residential smart microgrid. *IEEE Transactions on Power Electronics.* 2018;34(4):3283-301.
14. Al Badwawi R, Issa WR, Mallick TK, Abusara M. Supervisory control for power management of an islanded AC microgrid using a frequency signalling-based Fuzzy Logic Controller. *IEEE Transactions on Sustainable Energy.* 2018;10(1):94-104.
15. Jafari M, Malekjamshidi Z, Zhu J, Khooban MH. A Novel Predictive Fuzzy Logic-Based Energy Management System for Grid-connected and Off-grid Operation of Residential Smart Micro-grids. *IEEE Journal of Emerging and Selected Topics in Power Electronics.* 2018.
16. Prathyush M, Jasmin EA. Fuzzy Logic Based Energy Management System Design for AC Microgrid. *Second International Conference on Inventive Communication and Computational Technologies (ICICCT)*, IEEE. 2018:411-414.
17. Ji L, Zhang BB, Huang GH, Xie YL, Niu DX. Explicit cost-risk tradeoff for optimal energy management in CCHP microgrid system under fuzzy-risk preferences. *Energy Economics.* 2018;70:525-35.
18. Arcos-Aviles D, Pascual J, Guinjoan F, Marroyo L, Sanchis P, Marietta MP. Low complexity energy management strategy for grid profile smoothing of a residential grid-connected microgrid using generation and demand forecasting. *Applied Energy.* 2017;205:69-84.
19. Angalaeswari S, Swathika OG, Ananthakrishnan V, Daya JF, Jamuna K. Efficient power management of grid operated microgrid using fuzzy logic controller (FLC). *Energy Procedia.* 2017;117:268-74.
20. Yang L, Tai N, Fan C, Meng Y. Energy regulating and fluctuation stabilizing by air source heat pump and battery energy storage system in microgrid. *Renewable Energy.* 2016;95:202-12.
21. Moafi M, Marzband M, Savaghebi M, Guerrero JM. Energy management system based on fuzzy fractional order PID controller for transient stability improvement in microgrids with energy storage. *International Transactions on Electrical Energy Systems.* 2016;26(10):2087-106.
22. Chen YK, Wu YC, Song CC, Chen YS. Design and implementation of energy management system with fuzzy control for DC microgrid systems. *IEEE Transactions on Power Electronics.* 2012;28(4):1563-70.
23. Kyriakarakos G, Dounis AI, Arvanitis KG, Papadakis G. A fuzzy logic energy management system for polygeneration microgrids. *Renewable Energy.* 2012;41:315-27.
24. Chaouachi A, Kamel RM, Andoulsi R, Nagasaka K. Multiobjective intelligent energy management for a microgrid. *IEEE Transactions on Industrial Electronics.* 2012;60(4):1688-99.

25. Tabanjat A, Becherif M, Hissel D, Ramadan HS. Energy management hypothesis for hybrid power system of H2/WT/PV/GMT via AI techniques. *International Journal of Hydrogen Energy.* 2018;43(6):3527-41.
26. Diamantoulakis PD, Kapinas VM, Karagiannidis GK. Big data analytics for dynamic energy management in smart grids. *Big Data Research.* 2015;2(3):94-101.
27. Roy K, Mandal KK, Mandal AC. Ant-Lion Optimizer algorithm and recurrent neural network for energy management of micro grid connected system. *Energy.* 2019;167:402-16.
28. Ma YJ, Zhai MY. Day-Ahead Prediction of Microgrid Electricity Demand Using a Hybrid Artificial Intelligence Model. *Processes.* 2019;7(6):320.
29. Ghorbani S, Unland R, Shokouhandeh H, Kowalczyk R. An Innovative Stochastic Multi-Agent-Based Energy Management Approach for Microgrids Considering Uncertainties. *Inventions.* 2019;4(3):37.
30. Chui KT, Lytras MD, Visvizi A. Energy sustainability in smart cities: Artificial intelligence, smart monitoring, and optimization of energy consumption. *Energies.* 2018;11(11):2869
31. Ganesan S, Padmanaban S, Varadarajan R, Subramaniam U, Mihet-Popa L. Study and analysis of an intelligent microgrid energy management solution with distributed energy sources. *Energies.* 2017;10(9):1419.
32. Yan X, Abbes D, Francois B. Development of a tool for urban microgrid optimal energy planning and management. *Simulation Modelling Practice and Theory.* 2018;89:64-81.
33. Panbao WA, Wei WA, Nina ME, Dianguo XU. Multi-objective energy management system for DC microgrids based on the maximum membership degree principle. *Journal of Modern Power Systems and Clean Energy.* 2018;6(4):668-78.
34. Alhussein M, Haider SI, Aurangzeb K. Microgrid-level energy management approach based on short-term forecasting of wind speed and solar irradiance. *Energies.* 2019;12(8):1487.
35. Baghaee HR, Mirsalim M, Gharehpetian GB. Multi-objective optimal power management and sizing of a reliable wind/PV microgrid with hydrogen energy storage using MOPSO. *Journal of Intelligent & Fuzzy Systems.* 2017;32(3):1753-73.
36. Murugaperumal K, Raj PA. Energy storage based MG connected system for optimal management of energy: An ANFMDA technique. *International Journal of Hydrogen Energy.* 2019;44(16):7996-8010.
37. Sanseverino ER, Di Silvestre ML, Ippolito MG, De Paola A, Re GL. An execution, monitoring and replanning approach for optimal energy management in microgrids. *Energy.* 2011;36(5):3429-36.
38. Phan BC, Lai YC. Control Strategy of a Hybrid Renewable Energy System Based on Reinforcement Learning Approach for an Isolated Microgrid. *Applied Sciences.* 2019;9(19):4001.

39. Indragandhi V, Logesh R, Subramaniyaswamy V, Vijayakumar V, Siarry P, Uden L. Multi-objective optimization and energy management in renewable based AC/DC microgrid. *Computers & Electrical Engineering*. 2018;70:179-98.
40. Utkarsh K, Srinivasan D, Trivedi A, Zhang W, Reindl T. Distributed model-predictive real-time optimal operation of a network of smart microgrids. *IEEE Transactions on Smart Grid*. 2018;10(3):2833-45.
41. Bhattacharya A, Kharoufeh JP, Zeng B. Managing energy storage in microgrids: A multistage stochastic programming approach. *IEEE Transactions on Smart Grid*. 2016;9(1):483-96.
42. Yan B, Luh PB, Warner G, Zhang P. Operation and design optimization of microgrids with renewables. *IEEE Transactions on Automation Science and Engineering*. 2017;14(2):573-85.
43. Romero-Quete D, Cañizares CA. An affine arithmetic-based energy management system for isolated microgrids. *IEEE Transactions on Smart Grid*. 2018;10(3):2989-98.
44. Wang D, Qiu J, Reedman L, Meng K, Lai LL. Two-stage energy management for networked microgrids with high renewable penetration. *Applied Energy*. 2018;226:39-48.
45. Thirugnanam K, Kerk SK, Yuen C, Liu N, Zhang M. Energy management for renewable microgrid in reducing diesel generators usage with multiple types of battery. *IEEE Transactions on Industrial Electronics*. 2018;65(8):6772-86.
46. Netto RS, Ramalho GR, Bonatto BD, Carpinteiro OA, Zambroni de Souza AC, Oliveira DQ, Braga RA. Real-time framework for energy management system of a smart microgrid using multiagent systems. *Energies*. 2018;11(3):656.
47. Jafari M, Malekjamshidi Z, Zhu J. A magnetically coupled multi-port, multi-operation-mode micro-grid with a predictive dynamic programming-based energy management for residential applications. *International Journal of Electrical Power & Energy Systems*. 2019;104:784-96.
48. Nemati M, Braun M, Tenbohlen S. Optimization of unit commitment and economic dispatch in microgrids based on genetic algorithm and mixed integer linear programming. *Applied Energy*. 2018;210:944-63.
49. Cardoso G, Brouhard T, DeForest N, Wang D, Heleno M, Kotzur L. Battery aging in multi-energy microgrid design using mixed integer linear programming. *Applied Energy*. 2018;231:1059-69.
50. Vergara PP, López JC, da Silva LC, Rider MJ. Security-constrained optimal energy management system for three-phase residential microgrids. *Electric Power Systems Research*. 2017;146:371-82.
51. Tabar VS, Ghassemzadeh S, Tohidi S. Energy management in hybrid microgrid with considering multiple power market and real time demand response. *Energy*. 2019;174:10-23.
52. Marzband M, Sumper A, Domínguez-García JL, Gumara-Ferret R. Experimental validation of a real time energy management system for

microgrids in islanded mode using a local day-ahead electricity market and MINLP. *Energy Conversion and Management*. 2013;76:314-22.
53. Cheddadi Y, Gaga A, Errahimi F, Sbai NE. Design of an energy management system for an autonomous hybrid micro-grid based on Labview IDE. *International Renewable and Sustainable Energy Conference (IRSEC)*. IEEE. 2015:1-6.
54. Dabbaghjamanesh M, Kavousi-Fard A, Dong Z. A Novel Distributed Cloud-Fog Based Framework for Energy Management of Networked Microgrids. *IEEE Transactions on Power Systems*. 2020.
55. Nandipati SH, Babu PT, Chigurupati M, Vaithilingam C. Interface protection and energy management system for microgrid using internet of things. *Energy Procedia*. 2017;117:201-8.
56. Chen Y, Hu M. Balancing collective and individual interests in transactive energy management of interconnected micro-grid clusters. *Energy*. 2016;109:1075-85.
57. Mansour-lakouraj M, Shahabi M. Comprehensive analysis of risk-based energy management for dependent micro-grid under normal and emergency operations. *Energy*. 2019;171:928-43.
58. Shaterabadi M, Jirdehi MA. Multi-objective stochastic programming energy management for integrated INVELOX turbines in microgrids: A new type of turbines. *Renewable Energy*. 2020;145:2754-69.

8

Urban Solid Waste Management for Energy Generation

Shikha Patel[1]* and Reshmi Manikoth Kollarath[2]†

[1]*Department of Architecture and Urban Planning, Qatar University, Doha, Qatar*
[2]*BMS College of Architecture, Bangalore, India*

Abstract

This paper explores the authoritative knowledge of existing waste management practices and policies in Bangalore, India. The aim is to relate two critical issues: managing waste in large cities and converting it into energy once managed. The methodology allows a thorough literature review investigating the existing data and with the help of surveys and non-structured interviews with experts, identifying challenges of managing waste and transforming it into energy. The findings wind up in three scales, namely, city, neighbourhood, and household scales. The analysed data, shaped by an understanding of the challenges at these scales and drawn from the interviews and surveys' conclusions, is categorized into four themes: environmental, technological, social, and economic aspects. The results recommend policies to abate these challenges, promote growth, and foster a better transition towards more sustainable development.

Keywords: Waste management, waste-to-energy, Bangalore, challenges, policy analysis, urban planning, waste segregation, public awareness

8.1 Introduction

8.1.1 Background

Solid Waste Management is a major environmental issue across the world posing varied challenges both in developing and developed countries. Municipal Solid Waste (MSW) consists of both biodegradable and

---

*Corresponding author: sp2009946@qu.edu.qa; shikhamehta9281@gmail.com;
ORCID ID: 0000-0001-9130-1703
†*Corresponding author*: reshmimk@gmail.com; ORCID ID: 0000-0002-4707-4926

non-biodegradable wastes including food scraps, plastic bottles, furniture, clothing, paper, etc., which are sent to the landfill. Other than this, construction and demolition debris, non-hazardous industrial wastes and sludge from wastewater treatment are also sent to landfills even though they are not included in MSW. In recent years plastic and e-waste have become major contributors to the total waste being generated. The solid waste generated can be further categorized into domestic, agricultural, medical, and industrial [1].

Solid Waste Management has become a serious issue in recent years with a rapid rate of urbanization across the world and increasing population in the cities [2]. This leads to an alarming increase in the generation of waste which then gets dumped onto water bodies, roadsides, and vacant lands. The waste then pollutes the air, water and land, causing severe environmental damage.

Solid waste management is a major issue in all Indian cities with their high densities of population. According to the Ministry of Urban Development, Government of India, 2014, solid waste produced was 133,000 MT/day while total waste collected is 91,000 MT/day. Littered waste was 42,000 MT/day. 26,000MT/day of MSW collected is treated and 66,000 lakh MT/day is a land filled [3]. The Solid Waste Management Rules 2016, released by the Union Ministry of Environment, Forest and Climate Change give the necessary guidelines for waste disposal in India [4]. But compliance with this is very limited. In India, the informal sector plays an important role in waste recycling and generating value from waste. There are more than 1.6 million informal waste pickers in India [5]. Some of the key aspects mentioned in the Solid Waste Management rules include the segregation of waste into biodegradable waste, dry waste, and domestic hazardous waste at the source itself. Non-recyclable wastes with a calorific value of more than 1,500 Kcal/kg are not allowed in landfills. They must be used for generating energy or to prepare refuse-derived fuel [6].

Even though India is the second most populated country in the world, it only occupies 5% of the world's area. Uncontrolled urbanization and improved standards of living have contributed to increased waste generation in India. Compared to other developed countries India has a solid waste collection efficiency of only 70% [7]. Looking at the changing patterns of waste generation in India, the amount of biodegradable waste has fallen from 70-80% to 45-50% and other wastes, including plastic, paper, medical, and hazardous waste, have increased to 50% [8]. The untreated waste dumped on the roadsides affects the aesthetic beauty of the cities and has a foul smell. The waste contributes to several environmental issues including global warming and depletion of the ozone layer, and it also causes ecosystem destruction as well as health issues and diseases.

Recent technological developments have helped in generating energy from the waste while also reducing the amount of waste safely disposed of. Some of the methods used for the generation of energy include incineration, bio-methanation, pyrolysis and gasification [9]. Along with the generation of electricity, these methods help to reduce environmental pollution as well. The traditional waste management practices in India include composting, bio-methanation and landfilling. The first two processes are time-consuming and therefore are not enough to cater to the growing amount of waste being generated. Land Filling has become the most common method of solid waste disposal in India. But for this, India will require 1,240 hectares of land to be added every year for solid waste disposal [10, 11].

The main aim of waste-to-energy initiatives is to reduce the massive amounts of waste being generated. Even though the waste-to-energy sector was initiated in 1987, there are several reasons why it has not made many strides in this field. Some of the reasons are listed below:

   i) Poor quality of waste
  ii) High capital as well as high Operations & Maintenance Costs
 iii) Absence of indigenous technologies
  iv) Scarcity of successful projects in the sector
   v) Failure of ambitious projects
  vi) Absence of coordination between municipalities, state, and central government bodies
 vii) Largely dependent on government subsidies
viii) Difficulty in purchasing power agreements with state electricity boards
  ix) Not enough support from banks and other financial institutions

In recent years Swachh Bharat Mission has renewed the interest in waste in the energy sector as well.

8.1.2 Study Focus

Bangalore became the first city in India to start segregating waste at its source following a high court order which was a result of citizen-based activism in the city. With the lack of proper infrastructure facilities, waste dumping and waste disposal are major issues in the city. Improper waste management has also led to groundwater pollution because of the leachate percolating into the ground, which adds to the already existing water shortage in the city. As the city is rapidly urbanizing and expanding in all directions it has become important to find sustainable solutions to the Solid Waste Management issues facing the city.

The study aims to identify the ground practices and policies of waste management in India, especially in Bangalore. The challenges are categorized into various themes, namely, environmental, technological, economic, social, and political, for better understanding. The study suggests waste to energy as the key to meeting the identified challenges, recommending policies at city and neighbourhood scales to meet the challenges. The research is limited to proposing policies and strategies that are achievable and practical in the city of Bangalore. The previous successes and failures of waste-to-energy plants in India help to investigate and conclude the limitations of waste management in Indian cities.

8.2 Literature Review

Bangalore is the fifth-largest city in India with a population of 10.18 million. Bangalore Metropolitan Area covers 1258 sq. km. Bangalore is an inland city located at an elevation of 900 metres above sea level and enjoys a moderate climate.

Earlier known as Garden city, Bangalore is currently known as the Silicon Valley of India and has seen tremendous growth in recent years due to the growth of the IT industry in the city. Increasing job opportunities have resulted in large-scale migration into the city in the past decade. Presently the city generates 5,000 metric tons of waste per day. Per capita waste is estimated at 0.5 kg per day. With so much waste being generated, there are over 60 illegal waste dumps in the city [2].

Bruhat Bengaluru Mahanagara Palike (BBMP) is the agency responsible for solid waste management in the city. It has two departments that oversee MSWM. The Health Department is responsible for the collection and disposal of solid waste in the city while the Engineering Department is responsible for the collection of construction and other demolition wastes in the city [12]. The chart below shows the organizational structure of MSWM in Bengaluru.

A comparative study for the period 1997-2013 shows the change in waste composition in Bangalore [2]. There has been an increase in the percentage of biodegradable waste from 42% to 61% and a decrease in the paper, leather, and other recyclable wastes, as shown in Figure 8.1. This shows an increase in recycling happening in the city.

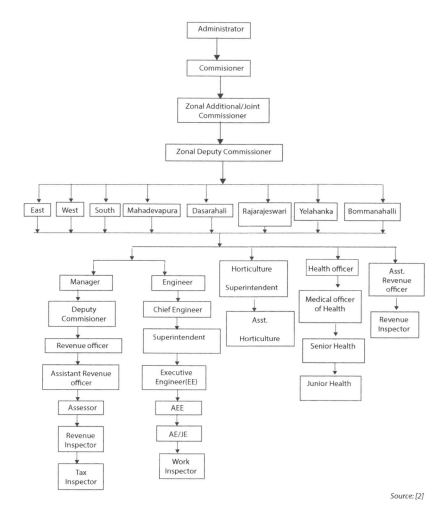

Figure 8.1 BBMP organizational chart.

Comparing data from studies done in 1999, 2000, 2005 and 2013, there is a substantial increase in organic waste generated, as shown in Figure 8.2. There is a decrease in other kinds of waste such as paper, leather, cardboard, rubber, glass, and metals. The percentage of plastic remains around the same. This clearly shows an increase in recycling as well as an increase in organic waste due to population increase and other urbanization pressures. There was a decrease in plastic waste between 2007 and 2013 which

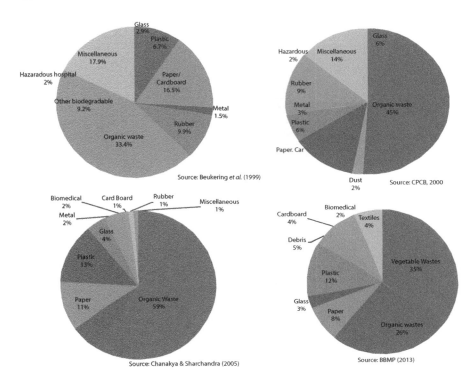

Figure 8.2 Waste composition in Bangalore.

could be because of the ban on plastic carry bags which was implemented in 2012.

As most of the waste generated in Bangalore is organic, composting has been a very successful technique that has been implemented in many small- and large-scale units. Organic wastes can also be used for biogas generation and the generation of electricity. The conventional methods used for organic waste disposal in Bangalore include composting, incineration, biogas generation and landfilling. But both composting and incineration produce further waste which must be then disposed of [13].

Currently, Bangalore can manage 2,100 TPD and achieve 100% waste management. The government has identified new landfill sites around Bangalore. But the dumping of waste in nearby villages is causing severe environmental damage in these villages and has also started polluting the wells and drinking water supply. The leachate from the waste contaminates the groundwater [14]. For instance, the landfill in Mavallipura village was closed in 2012 after the villagers protested after 10 lakh tons of waste was dumped there. The village still suffers from the environmental damage caused by the landfill and the groundwater and water bodies around the landfill area are still contaminated and black. It becomes imperative to find more sustainable solutions for wastewater treatment [15].

Some of the new technologies for treating organic waste include aerobic composting, windrow composting, community composting, vermicomposting, and biomechanical composting. Other than composting techniques several waste-to-energy initiatives can also be taken up which can also aid the energy issues in the city and neighbouring areas.

Waste-to-energy technologies that can be adopted include:

i) Biogas generation
ii) Refuse derived fuel
iii) Plasma gasification

Despite being one of the important cities, in terms of contribution to GDP, Bengaluru faces fundamental waste management issues. Lack of public motivation, which is partly because of lack of public awareness, plays an important role in households. In that regard, a household survey conducted for this study assisted the authors in understanding the issues and challenges that discourage people from making waste management a priority. From the interviews with government officials, it is observed that the responsibility also lies with the state and local bodies. One reason cannot be picked out for the failure of the waste management system in Bengaluru, rather it is a combination of systems at various scales and with several stakeholders.

8.3 Methodology

The scope of the study is to acknowledge the significance of waste-to-energy conversion, especially in high-consumption metropolitan cities such as Bengaluru. The research design is divided into three parts, as described below.

8.3.1 Formulating Research Background

The study adopts the mixed methodology of qualitative and quantitative data collection. Qualitative data comprises primary data obtained from desktop research on systems of waste management in India, in general. However, the study particularly focuses on waste management systems in Bengaluru intending to understand the scales, stakeholders involved and the challenges of managing waste. The study further investigates the presence of waste-to-energy conversion practices prevalent in Bengaluru and other Indian cities to identify its challenges.

8.3.2 Literature Review

The research background is followed by a thorough literature review extracted from refereed articles from journals, statutory reports, statistical reports, available policy data and information from government websites. The data is filtered based on the number of citations received by individual studies, the authenticity of the data and its relevance to this study. The secondary data comprises an analysis of results from survey respondents performed between 10th June 2021 and 25th June 2021. Figure 8.3 presents the structure of the study. In total, 100 respondents answered an online questionnaire, shared via social media. The targeted respondents were not limited to gender, race, ethnicity, social background, or place of residence in Bengaluru. The survey aimed to better understand people's willingness, awareness, and perception of the significance of managing waste

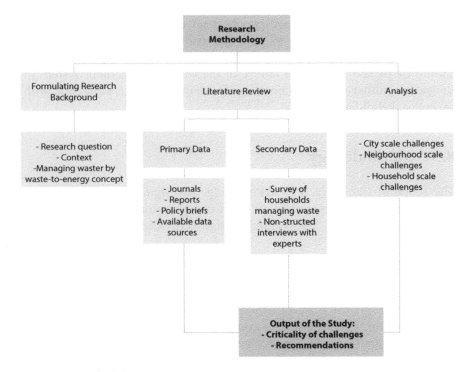

Figure 8.3 Methodology.

and waste-to-energy conversion. As a part of a non-structured interview, a focus group of experts from Bruhat Bengaluru Mahanagar Palike (BBMP) were targeted to understand the on-ground reality of challenges to managing waste.

8.3.3 Analysis

The secondary data collected in the form of surveys and interviews were further analysed to identify challenges of waste-to-energy conversion at various scales. For this study, the findings are presented at three scales, namely, city, neighbourhood and household, and are categorized into four themes: (1) environmental, (2) technological, (3) social, and (4) economic. In the final section, the authors recommend urban planning and public policy directives, hoping to amend the challenges of waste-to-energy conversion. The conclusions are based on a clear understanding of the collected surveys while the recommendations are based on the understanding of guidelines and policy directives planned for similar conditions.

8.4 Case Study

India probably has the highest population of activists and social workers involved in managing waste and the issue should have been resolved. But this is not the case. The efforts of activists go in vain because of the incompetent municipal system. Though there are a few success stories, there are relatively more failure stories to learn from. This section lists a few success and failure stories and unravels their causes.

8.4.1 Precedent Success

a) Waste-based power unit at Kathonda, Jabalpur (India)
This plant process 600 tonnes of waste per day generating 11.5 MW of electricity to provide to 18,000 households in Jabalpur. Operational since 2020, this project is currently running successfully under the PPP model. Though the waste is unsegregated, the latest technology used seems to produce power releasing minimum pollution in the end.

b) Modern garbage treatment park, Palampur, near Shimla (India)
A small panchayat of 7,000 residents, under the guidance of an NGO Shimla Environment, Heritage Conservation and Beautification Society (Sehab), collect segregated waste from households to be treated by separating at

Source: [16]

Figure 8.4 Non-segregated waste managing plant at Jabalpur.

the garbage treatment plant. There are about 800 volunteers or garbage collectors who make sure that the waste is segregated at the household level and only then taken to the plant as shown in Figure 8.4. The plant treats organic, inorganic, and plastic wastes separately. This is an ideal example of how waste should be managed; however, monitoring waste segregation in big cities with more than a million population is questionable [17].

8.4.2 Precedent Failure

a) Timarpur, New Delhi, India
The first large-scale plant to incinerate 300 tonnes of Municipal Solid Waste failed due to poor waste segregation methods, inefficient technology selection, variation in the quality of waste in different seasons and poor operations and maintenance [18].

b) Okhla waste-to-energy plant, New Delhi, India
The plant was established in 2007 to produce 16 MW of electricity by burning 600 – 720 tonnes of waste per day. However, the poor quality of waste, which is non-segregated waste with more than 50% moisture content, does not allow the plant to convert 100% of waste into energy. Consequently, the plant released toxic smoke from the unprocessed waste and created dangerous atmospheric pollution, which was unsafe and fatal for the nearby residential areas. The residents of South Delhi have filed a petition to shut

down the plant since 2009 and are still actively fighting to achieve their goal [19].

c) *Karimnagar plant, Telangana, India*
This plant was established in 2006 to process 1,400 tonnes of municipal waste per day generating 12 MW of electricity. The plant shut down in 2017 due to its inability to process poor-quality waste and expensive electricity which attracted unappealing power companies.

d) *Waste to green energy plant, Ramtekdi, Pune, India*
The plant was established in 2010 under the BOT model, planned to be run by Pune Municipal Corporation as shown in Figure 8.5. The plant was designed to process 700 tonnes of waste per day generating 10 MW of electricity. However, the plant proved to be highly inefficient and instead became a dump yard collecting more waste every day without processing it. The residents around the plant filed a petition in 2017 to shut it down as it created unhygienic surroundings for them. The decision is still to be made on shutting the plant down, as the builders of the plant argue that it was not designed to process as much waste as they receive every day [20].

Source: [21]

Figure 8.5 Waste to power plant turns to dump yard in Pune.

8.4.3 The Takeaway from Case Studies

The central government is actively tendering waste-to-energy plants under the PPP model. While there are a handful of success stories, more than 15 such big plants have shut down in the past decade [22]. Though waste-to-energy conversion has the scope and potential to solve waste management challenges in India, many plants have failed. The primary causes of the failure can be summarized into three issues: (1) poor quality of waste that is produced in the household. Despite growing awareness in many cities, the mass population still fails to understand the importance of waste segregation at the household level. (2) Installation, functioning and maintenance of waste-to-energy plants remain an expensive affair in India, despite decent subsidies granted by the government. Lack of efficient technology, lack of skilled labour, corruption, etc., can be pointed out as some of the reasons that it remains expensive. (3) The repeated violation of laws due to which public protests and petitions are triggered. It is established that the quality of waste is poor and there are no efficient and inexpensive processes to segregate waste at the waste-to-energy plant site, and as a result, the plant owners treat the waste poorly, creating toxic smoke unsafe for its surroundings. As the power created from these plants is subsidized at kilowatt per hour by the government, the owners tend to aim high on producing power, not caring about making the quality of waste better before processing it.

8.5 Research Findings: Challenges of Waste-to-Energy Conversion

According to a survey performed in 2015, the major and fundamental challenges of waste-to-energy conversion in a developing nation like India are poor waste segregation systems and the uptake of technological advancements [23]. Even though the Ministry of Urban Development is actively involved in managing Municipal Solid Waste (MSW) as a part of the Swachh Bharat Mission, the concept of waste-to-energy conversion is still young in India [24]. Consequently, the responsibility of managing waste and processing it into energy lies with urban local bodies (ULBs) in India, due to softly enforced laws at the national level.

8.5.1 Environmental Challenges

a) Overflowing landfills

One of the known and popular methods followed in India is collecting waste in open dumps and/or landfills. This is environmentally dangerous as it degrades expensive (and limited) urban land and discharges methane gas during the process of decomposition. Methane can not only cause urban fires but also is a major contributor to global greenhouse gases. Bangalore has recently adopted engineered landfills in neighbourhoods such as Kannahalli, Seegihalli, Doddabidarakallu, Lingaderenahalli, Subrayanpalya and Chikkanagamangala which significantly reduces the environmental impact of otherwise open landfills [2]. However, these plants only solve part of the problem of segregating compostable and non-compostable waste; the issue of converting it into usable energy remains. Poor-quality waste, which is non-segregated, remains the fundamental of all issues.

b) Water crises

During rainy days, these landfill's mountains overflow, and the waste is carried to the nearest streams and/or rivers. This unhygienic water mixed with landfill waste is used by city dwellers, making the health of people extremely vulnerable. In addition, many industries and neighbourhoods still discard their waste in nearby lakes and rivers, especially in the case of Bengaluru. Consequently, 85% of lakes (out of 392) are labelled as severely polluted and none of the lakes in Bengaluru is categorized under Grade A (drinking water) and Grade B (for bathing) water quality [25, 26]. The evidence of this is the Bellandur lake fire caused in 2015 due to an overflow of industrial and municipal waste.

c) Survey results

From the survey results, 82% of respondents said that they either discard waste in plastic bags or non-compostable material. This adds to the environmental challenges. Further enquiry revealed that the respondents face a lack of motivation and awareness to segregate waste and garbage into compostable bags. Also, 87% of respondents said they are not sure where their daily waste is taken to after they discard it into the neighbourhood garbage bin.

8.5.2 Technological Challenges

a) Incompetent machinery

A city as sprawling as Bengaluru, generating 5,000 metric tonnes of municipal solid waste each day, should have the most efficient and latest technology for converting waste to energy. But this is not the case due to all the reasons compiled in this paper: lack of political will, poor budgets allotted, ignorance of the citizens and environmentally damaging practices employed for decades. If waste segregation has failed at the household level, there must be a better and more efficient technology to segregate waste at the plant.

b) Lack of skilled staff

Along with delayed technological advancement in a developing nation like India, the availability of skilled labour and trained experts to work on waste-to-energy plants is also an important challenge [18]. Urban Local Bodies (ULB) fail to maintain and monitor waste-to-energy plants and/ or lack the expertise to do so. When the plants are designated to private companies to manage, that creates other financial liabilities for the ULBs.

c) Survey results

During an informal interview with a BBMP worker, it was revealed that the waste management sector faces an acute shortage of manpower. It tends to depend on volunteers and NGOs which are only seasonal workers. Due to the scale of the city and the amount of waste collected each day, it becomes unmanageable, and hence the easiest way is chosen, that is, dumping in landfills and lakes.

8.5.3 Social Challenges

a) Poor community participation

One of the major issues in India resulting from poor waste management practices is the lack of community participation. Examples from other cities around the globe have reported the benefits of public participation incentivized by local authorities. However, Bengaluru is much behind when it comes to community participation, primarily due to a lack of belongingness to the place. More than 50% of the population of urban Bengaluru has migrated from other cities and countries [27]. Due to the language barrier and general societal and lifestyle diversity, these migrants remain isolated from their communities. As a result, they develop apprehensive attitudes towards the welfare of the community [28]. This also leads to the issue of ownership of the waste.

b) Public health in danger
Residents, in the past, have reported serious health issues such as breathing issues, chronic diseases, reduced immunity, asthma and other infections caused by pollution of air around landfills and open waste dump yards [29]. In 2020, when BBMP proposed five waste-to-energy plants in Chikkanagamangala, the residents were uneasy and appealed to the local authorities. Their concerns are, (1) BBMP has not planned any segregation methods for a plant that is supposed to process only segregated waste, and (2) the plant is going to cause pollution and is located next to the neighbourhood. Residents are now worried that the 500 tonnes of waste brought to the site next to their homes and the consequent energy production on the plant are going to create air pollution along with other health issues.

c) Survey results
It is important to note that with waste-to-energy plants getting popular in Bengaluru, it raises issues of waste segregation, recycling and reuse. According to the survey results, residents presume that all the waste is anyway going to be burned and converted into electrical energy, so there remains no need to segregate [30]. However, the fact that unsegregated waste cannot be burned to 100%, creates a low-quality fuel and also leaves a bigger carbon footprint is ignored. Bengaluru needs a larger aware population and more activists to explain the importance of waste segregation. As per the survey responses, 5% of households compost their waste either at home or in their society. However, given the population of Bengaluru, this is a very small number to count.

8.5.4 Economic Challenges

a) Lack of available finances
Waste management is still not perceived as a primary issue among the many issues in Bengaluru. This leads to an apprehensive attitude from the state and local government regarding investing in managing waste and thereafter converting it into energy. Though the efforts from NGOs and activists are noticeable and inspiring, they have their constraints as far as making a change at the city level is concerned [31].

b) Little or no subsidies are provided for a waste collector

Collection of waste from streets and public areas is perceived as an informal activity as the waste pickers are not categorized under formal jobs. With no incentives, no job security and no support from municipal workers, waste pickers tend to use shortcuts, such as, instead of picking waste from streets and public places, they pick plastic waste from landfill sites and sell it to private entities. Most of these waste pickers are not even directly hired by the government, but are instead hired and paid by NGOs and private welfare bodies [32]. Hence, with no government support or socioeconomic security, these waste pickers tend to lose interest, which adds to the fundamental issue of waste collecting and segregation.

c) Survey results

According to the opinion of respondents, the government should give some soft incentives to the residents, such as a public forum where a locality can be given a waste-free title. Residents don't expect a monetary benefit only. As for the city scale, a bold rearrangement is needed to make waste management a priority. There are no programs or workshops to train ULBs in managing waste or converting it into energy, thereafter.

8.6 Recommendations

Recycling or converting waste to energy is not an overnight change that one can bring to communities. While Bengaluru is facing environmental, technological, social, and economic issues, it is not much different from other urban areas. The short-term goals and the smaller-level goals need community and individual participation. This will be the first step to achieving longer-term goals at the city, state, and national levels.

- For a multicultural metro city like Bengaluru, it is necessary to perceive waste management and waste-to-energy conversion as its priority among other city-level issues. This will help create awareness among people and encourage change in their attitudes.
- The private sector and NGOs need to pitch in to help ULBs to make people aware and incorporate public participation.
- Along with educating citizens on environmental degradation due to waste, it is necessary to educate citizens on the

dangers of not segregating waste. Based on the case studies, non-segregated waste is harmful when getting treated at a waste-to-energy plant. Campaigns from volunteers and NGOs should focus on dangers to public health.
- ULBs should lay a high focus on incentivizing communities and people's participation in an attempt to create awareness and responsibility among locals and migrants. This does not always have to be monetary benefits. ULBs can work on soft incentives as well, such as awarding the cleanest communities and neighbourhoods. This will not only encourage people but also can be beneficial to property owners in that locality.
- As a priority, open waste dumps and landfills should be replaced by properly designed and managed engineered landfills. This might not solve the entire environmental and public health issue but would reduce the impact of the waste.
- Municipal local bodies must invest in formalizing or at the least recognizing waste picker's jobs by giving them social security or job security. This small investment might help to solve the larger issue of waste segregation.
- There should be special programs and workshops designed for ULBs to get training in working at the waste-to-energy plant. If the state government takes charge of making their labour skilled, the system won't have to depend on the PPP model, which doesn't work for Indian cities. The BOT model might work much better if the ULBs are trained to run the plants.
- Instead of expensive city-level plants, which increase the hassle of collecting waste and transporting it to the plant site, the government can invest in plants at a neighbourhood scale. However, for this initiative, people's participation will be needed.

8.7 Conclusions and Discussion

The paper identifies the fundamental challenges faced by Indian cities, especially Bengaluru in managing waste. These challenges once resolved or given priority to resolve, are the first step towards converting waste to energy. The conclusions were drawn based on precedent success and failed plants across India and interviews with experts. The study also revealed the need to survey at the household level, as waste management must start with the smallest scale. The understanding is interpreted across four themes: environmental, technological, social, and economic. On the environmental level, the current practice of dumping waste into landfills proves to be highly unsustainable. Moreover, the unsegregated waste when sending to waste-to-energy plants creates more pollution due to the poor quality of waste and eventually disrupts the efficiency of the plant. On the technology level, the ULBs still suffer from poor finances from the state and central government to install city-level plants [33]. As a result, the city must subsidize electricity generated from the plant and install waste in the energy plant under the PPP model. This is found to be highly uneconomical and inefficient. The land, water and air pollution created by treating non-segregated waste-to-energy plants are hazardous to the citizens. And finally, one of the fundamental economic challenges is proved to be the inability of ULBs to pay and incentivize city waste pickers. Consequently, the waste pickers turn to unfair means to earn, which directly affects the quality of waste that is collected to be processed.

Through the suggestion at the city, neighbourhood and household levels, the authors hope to give direction to abate the challenges of waste management and contribute to making our cities more sustainable.

Acknowledgements

The authors are grateful to BBMP employees and residents of Bangalore for responding enthusiastically and positively to the surveys and questionaries.

References

1. United Nations ESCAP, "Introduction: Types of Waste," unescap.org, 2019.
2. N. BP and S. PV, "Solid Waste Management in Bengaluru - Current Scenario and Future Challenges," *Innovative Energy and Research*, vol. 5, no. 2, p. 139, 25 August 2016.

3. M. Kumar, S. Singh and S. Kumar, "Waste Management By Waste to Energy Initiatives in India," *International Journal of Sustainable Energy and Environmental Research,* vol. 10, no. 2, pp. 58-68, 21 April 2021.
4. S. S. Sambyal, Government notifies new solid waste management rules, downtoearth.org.in, 2016.
5. H. Yang, M. Ma, J. R. Thompson and R. J. Flower, "Waste management, informal recycling, environmental pollution and public health," *National Natural Science Foundation of China ,* pp. 1-20, 2019.
6. Ministry of Environment, Forest and Climate Change, "The Solid Waste Management Rules," Government of India, New Delhi, 2016.
7. M. Sharholy, "Municipal solid waste management in Indian cities – A review," *ScienceDirect,* pp. 459-467, 2008.
8. Y. B. P. e. a. Nandan A, "Recent Scenario of Solid Waste Management in India," *World Scientific News,* pp. 56-74, 2017.
9. Energy Saving Trust, Generating energy from waste: how it works, https://energysavingtrust.org.uk/generating-energy-waste-how-it-works/, 2020.
10. I. J. Ahluwalia and U. Patel, "Solid Waste Management in India An Assessment of Resource Recovery and Environmental Impact," Rockefeller Foundation, 2018.
11. R. Ilangovan, Will India Need a Landfill the Size of Bengaluru By 2030?, thewire.in, 2017.
12. Ministry of Urban Development, "Municipal Solid Waste Management Manual," Government of India, 2016.
13. M. S. Ayilara, O. S. Olanrewaju, O. O. Babalola and O. Odeyemi, "Waste Management through Composting: Challenges and Potentials," *Sustainability,* vol. 12, pp. 4456, doi:10.3390/su12114456, 30 May 2020.
14. A. M, Bengaluru dumping its garbage in nearby villages, Bengaluru: https://economictimes.indiatimes.com/news/politics-and-nation/bengaluru-dumping-its-garbage-in-nearby-villages/articleshow/70599664.cms?-from=mdr, 2019.
15. A. Sen, How Bengaluru's garbage killed Mavallipura's environment, Bengaluru: https://bengaluru.citizenmatters.in/mavalipura-landfill-problems-bangalore-garbage-dump-ecology-29749, 2018.
16. M. Tan, Circulate Capital to invest US$19m in four India waste management firms, https://www.businesstimes.com.sg/asean-business/circulate-capital-to-invest-us19m-in-four-india-waste-management-firms, 2020.
17. A. Sharma, How a small Himalayan town panchayat is showing Shimla the way to effective waste management, Shimla: https://citizenmatters.in/himachal-mountain-towns-shimla-palampur-waste-to-energy-plant-14354, 2019.

18. S. Kumar, S. R. Smith, G. Fowler, C. Velis, S. J. Kumar, S. Arya, Rena, R. Kumar and C. Cheeseman, "Challenges and opportunities associated with waste management in India," *Royal Society Open Science*, pp. 1-11, http://dx.doi.org/10.1098/rsos.160764, 22 March 2017.
19. R. Agarwal, Okhla waste-energy plant gets show cause notice, Delhi: https://www.downtoearth.org.in/news/waste/okhla-waste-energy-plant-gets-show-cause-notice-64108, 2019.
20. V. Chavan, Waste-to-power plant turns into dumping yard, Pune: https://punemirror.indiatimes.com/pune/civic/waste-to-power-plant-turns-into-dumping-yard/articleshow/59906789.cms, 2017.
21. V. Chavan, Hadapsar is next PMC dump yard, https://punemirror.indiatimes.com/pune/civic/hadapsar-is-next-pmc-dump-yard/articleshow/78736821.cms?utm_source=contentofinterest&utm_medium=text&utm_campaign=cppst, 2020.
22. S. S. Sambyal, R. Agarwal and R. Shrivastav, Trash-fired power plants wasted in India, Delhi: https://www.downtoearth.org.in/news/waste/trash-fired-power-plants-wasted-in-india-63984, 2019.
23. J. D. Nixon, D. Wright, J. A. Scott and P. K. Dey, "Issues and Challenges of Implementing Waste-to-Energy Practices in India," *Renewable Energy in the Service of Mankind*, vol. 1, pp. 64 - 75, DOI:10.1007/978-3-319-17777-9_7, 10 September 2015.
24. M. Muthuraman, "Waste to Energy - Challenges & Opportunities in India," in *National Seminar on Emerging energy scenario in India - Issues, Challenges and way forward*, India, 2018.
25. S. Goswami, About 85 per cent of Bengaluru's water bodies severely polluted: study, Bengaluru: Down to Earth, 2017.
26. K. Ritter, Booming Infrastructure Poisons Bangalore's Lakes, Depletes Groundwater, Bengaluru: Circle of Blue, 2018.
27. The Times of India, Bengaluru's migrants cross 50% of the city's population, Bengaluru: https://timesofindia.indiatimes.com/city/bengaluru/bengalurus-migrants-cross-50-of-the-citys-population/articleshow/70518536.cms, 2019.
28. M. Price and E. Chacko, "Understanding the issue of migrant exclusion," in *Migrants' Inclusion in Cities: Innovative Urban Policies and Practices*, 2012.
29. Ministry of Environment, Forest and Climate Change, "Central Pollution Control Board," 26 July 2017. [Online]. Available: https://cpcb.nic.in/municipal-solid-waste/. [Accessed 23 June 2021].
30. S. Chatterjee, Activists are opposing a waste to energy plant in Bengaluru: Here's why, Bengaluru: The News Minute, 2019.
31. K. Muniyappa and N. N., "Community attitude, perception and willingness towards solid waste management in Bangalore city, Karnataka, India," *International Journal of Environmental Sciences*, vol. 4, no. 1, pp. 87 - 95, July 2013.

32. S. Kanekal, "Challenges in the Informal Waste Sector: Bangalore, India," Penn Institute of Urban Research, Bengaluru, November 2019.
33. e. a. Sharholy M., "Municipal solid waste management in Indian cities – A review," *ScienceDirect,* pp. 459-467, 2008.

9

Energy from Urban Waste: A Mysterious Opportunity for Energy Generation Potential

Shivangini Sharma and Ashutosh Tripathi*

Amity Institute of Environmental Sciences, Amity University Uttar Pradesh (AUUP), Gautam Buddha Nagar, Uttar Pradesh, India

Abstract

Solid waste management has been one of the most pressing issues faced by humanity, especially in urban areas. The modern-day lifestyle has led to increased waste production. Population explosion and migration to urban areas is the reason behind failed management practices. Before the concept of "engineered landfills" came into the picture, heaps of municipal solid waste had already been created in developing countries. These landfill sites have long been considered overburdened and overloaded, but the dumping of waste continues. One of the most practical concepts for dealing with this waste is its utilization for obtaining energy. Various advanced technologies such as methane capture technology, plasma-pyrolysis, gasification, incineration, bio-methanation, etc., have the capability to extract useful energy from this legacy waste. All these processes work on different components of municipal solid waste. One of the most important aspects of treating municipal solid waste is the type of pollutants it releases during these processes. It is a matter of research to curb this sort of pollution. This chapter talks about how the global, as well as local scenario of municipal solid waste management, has altered over time. Also, there is a discussion about the various processes and technologies which can yield energy from waste. In the last portion of the chapter, various gaps in the implementation of these process technologies have been discussed. It can be concluded that efficient solid waste management can be achieved by the application of "waste-to-energy" and "circular economy" concepts.

Keywords: Waste to energy, biofuel, solid waste, SDG

---

*Corresponding author: atripathi1@amity.edu; ORCID ID- 0000-0003-4469-7017

9.1 Introduction

Discussion on waste generation and management has been an integral part of almost every global meeting today. The UN's Sustainable Development Goals (SDGs) nos. 12 and 13 are directly related to it. SDG 12 talks about sustainably reducing waste generation through prevention, reuse, reduction, and recycling. Also, SDG 7 can be connected to this. SDG 7 is about ensuring access to affordable, reliable, sustainable, and modern energy for all. This implies that, by 2030, the share of renewable energy will increase substantially [1].

The two most important reasons for increased waste generation are population growth and the fast pace of urbanization. Inefficient disposal of untreated waste on land creates several issues which have been ever-increasing. A lot of efforts are being made worldwide to address this issue effectively. Another growing concern for humans today is the fast depletion of conventional energy resources. Humans have been completely reliant on non-renewables for their energy needs for centuries and thus eventually over-exploitation has resulted in a decline in the available deposits. This not only has been a rising concern today but will be affecting future generations in the long run. In addition to the growing concerns about the diminishing resources of non-renewable fuels, the amount of pollution caused by these fuels is also far from what can be handled by nature. In the words of the executive director of International Energy Agency (IEA) Dr. Faith Birol, "Global carbon emissions are set to jump by 1.5 billion tonnes this year. This is a dire warning that the economic recovery from the Covid crisis is currently anything but sustainable for our climate" [2] (IEA, 2021).

India is responsible for nearly 6.65% of total global carbon emissions and is ranked fourth, behind China (26.83%), the USA (14.36%), and the EU (9.66%). Most of the power that is being generated in India is from coal and mineral resources. Such resources contribute to greenhouse gas emissions heavily. Coal, petroleum, and natural gas contribute about one-third of the global greenhouse gas emissions. Among the largest consumers of coal in the world, India lies in the top five and imports fossil fuels of high cost [3]. About 74% of its energy demand is fulfilled by coal and oil. A report from the Centre for Monitoring Indian Economy states that the country's import of coal has gone up from 171 million tons of coal in 2013–2014 to 213 million tons in 2017–2018 [4]. From 1960 to 2021 the population of India increased from 450.55 m to 1.39 bn people. And the amount of waste produced is 0.2 Kg to 0.6 Kg per day per capita in cities with a population ranging from 1.0 lakh to 50 lakh (Guidelines, 12th Finance Commission Grants). Managing this much waste has been a daunting task for the authorities and despite various efforts, there still is a huge gap between the formulation and execution of various plans.

Managing waste takes about 20-50% of the total municipal budget as per the World Bank [9]. Plenty of technologies have already been proposed but none is a completely stable solution addressing the entire problem at once. Thus, there has begun a hunt for newer technologies for generating energy from renewable resources such as solar, wind, biomass, etc. The last 25 years have seen a phase of intensive research and developmental activities in India for finding out alternative renewable energy resources for compensating with the needs of the rapidly growing population of the country.

9.2 Scenario of Solid Waste Management of Various Countries Around the World

The generation of municipal solid waste in Asia had an annual growth rate of around 3% in developed countries and around 5% in developing countries in the 1990s. A lot of initiatives for solid waste management began around this decade due to the foreseen challenge of pollution load. Some of these initiatives include the Metropolitan Environmental Improvement Program by the World Bank for the management of municipal solid waste in large metropolitan cities such as Bombay, Beijing, Colombo, etc. From 1994 to 1998 the Canadian International Development Agency (CIDA) successfully ran a program for MSW management named.

The South East Asia Local Solid Waste Improvement Project in the Philipinnes, Indonesia and Thailand. Under this program, a waste bank was set up for managing recyclables, various training programmes were initiated for waste handling and landfill management also gained pace [5]. At present these densely populated cities are under pressure of upgrading their solid waste management systems by mainstreaming newer technologies. The idea of a so-called sanitary landfill dates back to the 1950s in the U.S. when the waste was put in thin layers that were compacted on top of each other to reduce its volume and put a layer of soil on top. But with the establishment of the Environmental Protection Agency in 1970, the management of municipal solid waste started becoming more of an organized task. Within 2-3 decades, engineered landfills came into existence and by 2013, 87.2 million tonnes (15 million

144 URBAN ENERGY SYSTEMS

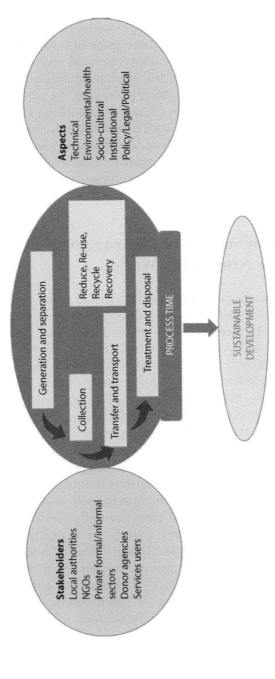

Figure 9.1 Diagrammatic concept representation of integrated sustainable solid waste management (ISSWM).

tonnes in 1980) of material was recycled and composted, saving it from being simply dumped. The idea of "integrated sustainable solid waste management" came from a Dutch NGO called WASTE in the mid-1980s [6]. Figure 9.1 shows a diagrammatic representation of ISWM strategies in short.

In European countries, the landfill diversion aimed at handling biodegradable municipal solid waste were given in Landfill Directives. Recycling targets given in the Packaging and Packaging Waste Directives and the Waste Framework Directives are the most relevant policies. An increase from 22% to 29% in the waste recycling rates was seen from 2004 to 2012 following these policies [7].

Regarding the present-day estimates of waste generation around the world, the total solid waste generated in 2020 was about 2.4 billion tonnes or 0.79 kg of waste per person per day. By the year 2050, it has been estimated that this figure would rise by 73% to become 3.88 billion tonnes. This rise is attributed to growth in urbanization and growth in incomes of various countries in terms of GDP [8].

According to the report of UNSEDA, 2010, the GDP of the U.S. was found to be the highest but also was the MSW generated globally [9]. The World Bank's "What a Waste" program is a global initiative providing a database for monitoring waste generation and management across various countries of the world. What a Waste is a publicly accessible database initiated in 2018 containing statistics related to various aspects such as disposal, generation, collection, etc.. This database covers more than 330 cities around the world. Total waste generation by 2050 is going to increase about three and two times compared to the figures for 2016 in SSA (Sub-Saharan Africa) and SA (South Asia) due to economic growth. The Figure 9.2 shows that maximum waste generation in the long-term and short-term periods will be from these areas [10].

The growth in the urbanization of the already present population will only lead to a growth in the quantity of waste generated and the complexity of waste. Since in urban areas, the land is a limited resource, and its judicious use is very important, mindless dumping of municipal solid waste on huge patches of land is not favourable. Many technologies are therefore in the frame today which are helping better MSW management in both developing and developed countries.

Waste treatment around the world has seen a rapid evolution. Waste treatment pertains to the treatment given to waste lying in landfills. Scientific waste burning as well as composting are the two famous methods for legacy waste management. Source reduction and reuse, however, are based on the management of waste at the level of the producer

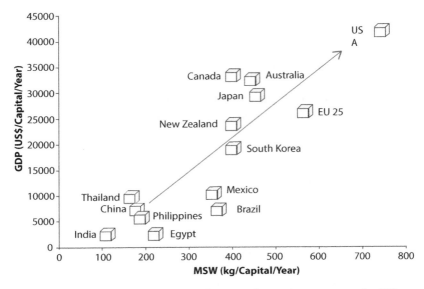

Figure 9.2 Growth in GDP and corresponding growth in MSW generation for different countries [9].

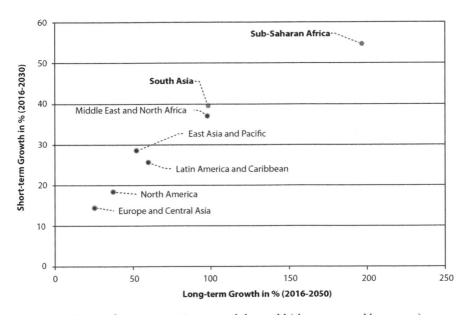

Figure 9.3 Regional waste generation around the world (short-term and long-term) (Source: [10]).

itself. The extraction of energy from the waste lying in landfills has various aspects, which will be discussed henceforth in the chapter. Figure 9.2 shows a diagram for growth in GDP and corresponding growth in MSW generation for different countries. Figure 9.3 shows a diagram for regional waste generation around the world (short-term and long-term).

9.3 Waste-to-Energy Processes

Waste has a huge potential to be converted into energy and with appropriate technology, it is possible to harness this energy.

All three types of processes are explained in detail in the following sections.

1. Thermo-chemical processes for waste-to-energy conversion
These processes, as the name suggests, involving the conversion of waste into energy using heat and various chemical treatments. Thermochemical processes usually include incineration, gasification, and pyrolysis.

a) Incineration
This is the process of burning mass waste at high temperatures, usually 1000 degrees C in the presence of air. The residues left after burning the waste in incineration furnaces include bottom ash, fly ash, metals and various air pollutants depending on the constituents of the fed waste. The outputs of the incineration process include heat and electricity. Incineration is the most widely used technology around the world for waste management. Usually, the incineration plants are coupled with flue gas capture and treatment systems and the bottom ash and fly ash collected are also used for various purposes. Bottom ash is usually unburnt or partially burnt waste material which gets collected at the bottom of the furnace due to its weight. Fly ash on the other hand is lighter and flies up in the furnace and is collected using various filter systems such as baghouse filters (made of cloth-like fibrous membranes), cyclonic precipitators, electrostatic precipitators, etc. This fly ash has variable amounts of various metals, metal oxides, silicon oxides, etc., depending on the waste characteristics. Due to high silica content, calcium oxides and aluminium oxides, etc., fly ash is often seen to have pozzolanic properties.

Hence, it is mixed with common Portland cement for providing extra strength. Due to the presence of harmful heavy metals, it cannot be used as a soil conditioner or fertiliser. Air pollutants expected from the incineration of common urban municipal waste include dioxins, particulate matter, Sox, NOx, carbon dioxide, etc. The gaseous pollutants are often treated with wet and dry scrubbers and particulate matter using fibrous filters. Following is the diagram depicting the common incineration process [11]. Figure 9.4 shows

148   URBAN ENERGY SYSTEMS

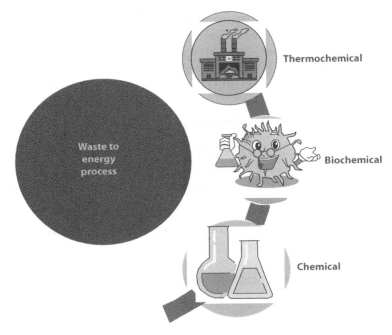

Figure 9.4 Waste-to-energy processes can be divided widely into three categories, Thermochemical, Biochemical and Chemical processes.

a diagram for waste-to-energy processes can be divided widely into three categories, thermochemical, biochemical and chemical processes.

b) Gasification
This is the process of burning waste in the presence of air at a temperature of around 750 degrees C. But when plasma arc is used for ignition, temperatures as high as 4000-12000 degrees C are reached. In this process, flue gases and bottom ash get produced. But the final useful gases formed are hydrogen, carbon monoxide, carbon dioxide, syngas, or producer gas, etc. These gases are used for the generation of electricity. The pollutants released after the gasification of municipal solid waste include tar, gas pollutants, inert gases, and hydrocarbons [13].

Gasification is often of two types, direct gasification, and indirect gasification. Direct gasification is usually partially aided by a gasification agent while in indirect gasification, no gasification agent is required. During direct gasification, an oxidation reaction occurs (oxygen/air is used) which supplies the required energy, while during indirect gasification reaction, steam is used which increases the amount of producer gas generated. While gasification is done using pure oxygen and has certain advantages in the production of a

Energy from Urban Waste 149

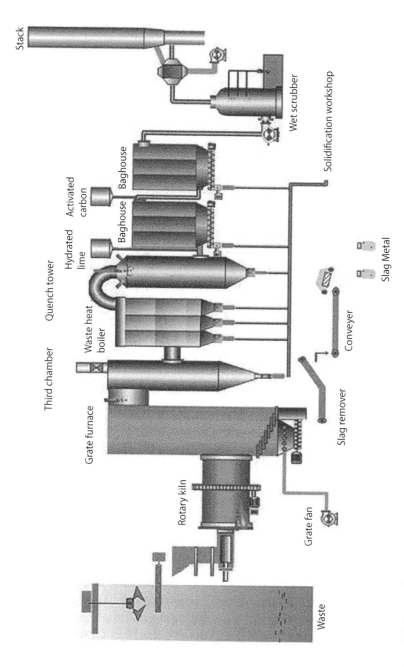

Figure 9.5 Setup of an incineration plant [12].

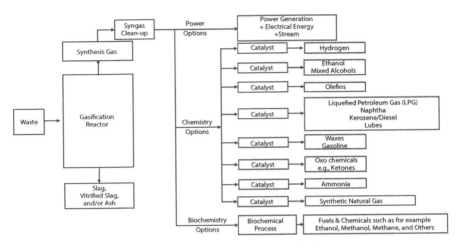

Figure 9.6 Diagrammatic view of the gasification process [14].

higher amount of energy and complete combustion, pure oxygen is expensive and the whole process becomes economically non-feasible. Figure 9.5 shows a setup of an incineration plant. Figure 9.6 shows a diagrammatic view of the gasification process.

c) Pyrolysis

It is a process involving heating any solid or liquid to a temperature of 400-1000 degrees C in the absence of oxygen and very high pressure. This anaerobic process results in a rapid degradation/separation of complex compounds into their simple constituents. Due to the absence of air/oxygen, during pyrolysis, combustion does not occur. Pyrolysis of municipal solid waste samples involves the breakdown of organic fractions of waste into products like solid carbon char, hydrogen gas and syngas. Along with the useful products, air pollutants are also released after the process of pyrolysis. After the pyrolysis of organic matter, some of the combustible gases can be easily condensed into combustible liquids. Some examples of such liquids include bio-oils which are collectively called pyrolysis oils. Thus, it can be said that almost all the products of pyrolysis of organic fraction of waste can be of economic value. Solid biochar can be used to produce membrane filter systems, bio-oils can be used as liquid fuels and syngas can be used for various other chemical reactions. Pyrolysis may be of various types depending on the solid/liquid material being treated or the chemical reaction taking place in the process [15]. Figure 9.7 shows a schematic of pyrolysis process to yield various products.

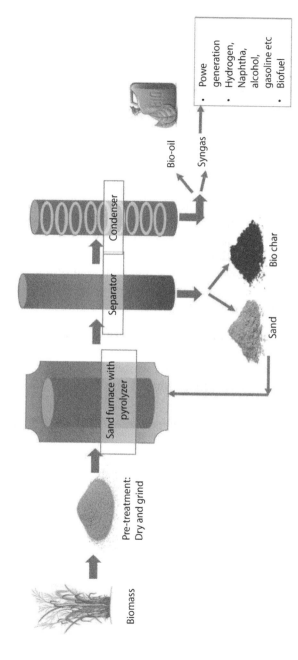

Figure 9.7 Schematic of pyrolysis process to yield various products.

2. Bio-chemical processes for waste-to-energy conversion

Biochemical processes involve the treatment of waste with microbes to obtain desired products in the form of energy, compost, gases, etc.

Fermentation is a biochemical process in which organic waste rich in sugar residues can be treated with anaerobic bacteria to decompose it to get energy alcohol, hydrogen gas and hence energy. The pollutants which are expected to be formed during this process may include leachates with microbial residues etc.

Methane gas captured from landfill waste can be used for the generation of electricity. Methane can be generated due to the decomposition of the organic portion of waste. This decomposition may be either naturally carried out by already present microbes or by intentional composting of waste by the addition of microbes. A microbial fuel cell is a setup in which microbes are used for the decomposition of organic waste to emit methane gas which is then converted into electricity. The residues of such processes include carbon dioxide, water, hydrogen gas, etc. Landfill gas capture is a very effective way of harnessing energy from waste, but it is a bit on the higher end as far as cost-effectiveness is concerned [16].

3. Chemical methods for generation of energy from waste

Esterification is the process of production of ester from acid and alcohol in the presence of an acid catalyst. The waste type used for this process is generally waste oil and the useful product is biodiesel and hence electricity.

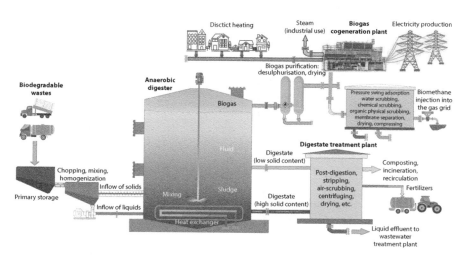

Figure 9.8 Diagrammatic view of anaerobic digestion of MSW [16].

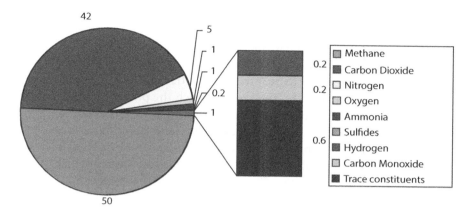

Figure 9.9 Percentage composition of typical landfill gas [18].

Wastewater is also a product of this reaction. The drawback of this method is its expensive nature [17]. Figure 9.8 shows a diagrammatic view of anaerobic digestion of MSW. Figure 9.9 shows a percentage composition of typical landfill gas.

$$R_1-\overset{O}{\underset{\|}{C}}-OH + R_2-OH \xrightarrow{Acid} R_1-\overset{O}{\underset{\|}{C}}-O-R_2$$

Carboxylic acid Alcohol → Ester + H$_2$O Water

9.4 Challenges to Waste-to-Energy Generation

1. Source separation: Most developing countries fail at the source separation stage of management. The main cause of the failure of source separation could be a lack of awareness among the generators. Due to this reason, many waste-to-energy plants have failed such as the bio-methanation plant based in Lucknow, India.
2. Waste composition: waste generated in developed countries is very distinct in its composition from that of developing countries. In developing countries like India, there is usually high moisture content and low organic content in the waste, resulting in low calorific value of the waste. This decreases the efficiency of the waste-to-energy plants.

3. Suitable technology: Most of the developed countries like the U.S.A., China, Japan, etc., have effectively managed their waste with appropriate technology available to them. On the other hand, most developing countries such as South Africa still lack appropriate waste conversion technologies.
4. Appropriate policies: Various strategies formulated on paper have not been able to bear fruit in practice due to the absence of a basic field study. The area in which the policy would be implemented should be studied thoroughly. For example, Municipal Solid Waste Management (and handling) rules were amended recently in 2016 in India. These rules still have not been able to be fruitful in managing waste because of the lack of surveys and research backing the execution.
5. Funds: There has always been an issue of lack of proper financial assistance from authorities in the waste management sector. For example, recent technological developments have made possible methane capture from engineered landfills, but in India, even in the capital city of Delhi, this has not been achieved. Financial assistance is required for the capping of landfills and the development of proper drainage systems for leachate.
6. Private invasion: It is very little to no invasion of the private sector in the waste management framework. Even in India, the calorific value of the waste is not up to the expectations of the industries and hence, its purchase is facing issues.
7. People's participation: Although many NGOs are working towards plastic recycling in India, there is still a lack of participation at the ground level. At source reduction, recycling and segregation by households would be extremely helpful in further management. But due to a lack of interest and knowledge, people do not coordinate with the authorities.

9.5 Conclusion

Energy waste is a promising concept and is already bearing fruit in many developed countries around the world. Countries like China, Japan, Denmark, etc., have already started extracting useful value-added products as well as energy from municipal solid waste using appropriate technologies. On the other hand, developing countries still lag due to various reasons as mentioned above. The road to obtaining energy from waste has many additional opportunities for developing countries. Employment generation, increase in household incomes, better health and sanitation, etc., are a few of them. Legacy waste lying in open unengineered landfills is the biggest hurdle in the path of growth and development. Thus, research should focus on finding better solutions for the treatment of legacy waste since the waste collected presently can easily be cut down in amount at source by educating the people of the nation.

References

1. Velenturf, A. P. M., & Purnell, P. Resource recovery from waste: Restoring the balance between resource scarcity and waste overload. *Sustainability (Switzerland)*, 2017;9(9).
2. Pappas, D., & Chalvatzis, K. J. Energy and Industrial Growth in India: The Next Emissions Superpower? In: *Energy Procedia, 105*. 2017.
3. Wec. World Energy Scenarios: Composing energy futures to 2050. Report From http://www.worldenergy.org/. 2013.
4. Blondeel, M., & van de Graaf, T. Toward a global coal mining moratorium? A comparative analysis of coal mining policies in the USA, China, India and Australia. *Climatic Change*, 2018;150(1–2).
5. Bag, S. Solid Waste to Energy Status in India: A Short Review. *Engineering and Technology Journal*. 2017. Nov 28.
6. Sharma, N.K., Sharma, S. Municipal solid waste management in developing countries: Future challenges and possible opportunities. *Journal of Green Engineering*, 2020;10(10).
7. Sharma, K.D., Jain, S. Municipal solid waste generation, composition, and management: the global scenario. *Social Responsibility Journal*, 2020;16(6).
8. Becker, F.G., Cleary, M., Team, R. et al. A global snapshot of solid waste management to 2050. In *Syria Studies* (Vol. 7, Issue 1), 2015.
9. United Nations. Report. World Economic Situations And Prospects. 2010.
10. Kaza, S., Yao, L.C., Bhada-Tata, P., van Woerden, F. What a Waste 2.0: A Global Snapshot of Solid Waste Management to 2050. In *What a Waste 2.0: A Global Snapshot of Solid Waste Management to 2050*. 2018.
11. Jeswani, H. K., & A. Incineration of Municipal Solid Waste. *Waste Management*. 2016.
12. Jiang X, Li Y, Yan J. Hazardous waste incineration in a rotary kiln: a review. Vol. 1, Waste Disposal and Sustainable Energy. Springer; 2019. p. 3–37.
13. Hooshmand, P., KhakRah, H.R., Balootaki, H.K., Abdollahzadeh Jamalabadi, M. Y. Recycling municipal solid waste utilizing gasification technology: a case study. *Journal of Thermal Analysis and Calorimetry*, 2020;139(4).
14. Seo, Y.-C., Alam, M.T., Yang, W.-S. Gasification of Municipal Solid Waste. In *Gasification for Low-grade Feedstock*. 2018.

15. Li, Q., Faramarzi, A., Zhang, S., Wang, Y., Hu, X., Gholizadeh, M. Progress in catalytic pyrolysis of municipal solid waste. In *Energy Conversion and Management* (Vol. 226). 2020.
16. Zupančič, M., Možic, V., Može, M., Cimerman, F., Golobič, I. Current Status and Review of Waste-to-Biogas Conversion for Selected European Countries and Worldwide. In *Sustainability (Switzerland)* (Vol. 14, Issue 3). 2022.
17. Esan, A.O., Olalere, O.A., Gan, C.Y., Smith, S.M., Ganesan, S. Synthesis of biodiesel from waste palm fatty acid distillate (PFAD) and dimethyl carbonate (DMC) via Taguchi optimisation method. *Biomass and Bioenergy*, 2021; 154.
18. Tchobanoglous, G., Kreith, F., Williams, M.E. Integrated Solid Waste Management Engineering Principles and Management Issues. In *Environmental Pollutants and Their Bioremediation Approaches*. 1993.

10

Sustainable Urban Planning and Sprawl Assessment Using Shannon's Entropy Model for Energy Management

Pranaya Diwate[1], Priyanka Patil[2], Pranali Kathe[2] and Varun Narayan Mishra[3]*

[1]*University Department of Basic and Applied Sciences, MGM University, Aurangabad, India*
[2]*Centre for Climate Change and Water Research, Suresh Gyan Vihar University, Jaipur, India*
[3]*Amity Institute of Geoinformatics & Remote Sensing (AIGIRS), Amity University Uttar Pradesh (AUUP), Gautam Buddha Nagar, Uttar Pradesh, India*

Abstract

Land use/land cover (LULC) changes are one of the most crucial factors for analyzing the unprecedented rate of urban sprawl. Most regions around the world are facing negative consequences due to rapid growth in urban areas and population. Thus, the urban sprawl assessment is vital for different planning and management activities. Understanding the process and patterns of urbanization will be helpful for the current and future development scenarios of a region. This will also further be beneficial for different problem-solving purposes, i.e., to know the availability and potential of solar energy resources. In the present study, temporal remote sensing datasets of the years 1990, 2000, 2010, and 2020, along with secondary data, were used for detecting LULC changes in Gandhinagar district, India. Supervised classification of the images was conducted to derive LULC maps. The total classification accuracies were found to be around 80%, 87%, 86%, and 80% for classified maps of 1990, 2000, 2010 and 2020, respectively. The results show the built-up land increased by 42.70%, 44.64%, and 18.51% during 1990-2000, 2000-2010, and 2010-2020, respectively. Shannon's entropy model has developed an understanding of the present and projects the future scenario of urban growth. The area under investigation saw an increase in built-up land at the expense of loss in the vegetated surface. The present study emphasizes the value of using remote

---

*Corresponding author: varun9686@gmail.com; https://orcid.org/0000-0002-4336-8038

Deepak Kumar (ed.) Urban Energy Systems: Modeling and Simulation for Smart Cities, (157–170) © 2023 Scrivener Publishing LLC

sensing to examine the kind and extent of ongoing changes. The study region saw a quick increase in the number of buildings at the expense of the natural cover. This work also emphasizes the use of remote sensing pictures in creating efficient master plans and management for regulated urban expansion at both the regional and local levels.

Keywords: LULC, urbanization, Shannon entropy, remote sensing, Landsat

10.1 Introduction

Cities around the world are experiencing greater urbanization challenges, mostly in developing countries [1–3]. The rapid population growth remarkably changed urban land use at different scales [4–6]. Evaluation of urban land-use changes and its sprawl plays a key role in achieving a city's sustainable planning strategies. Additionally, urbanization has several detrimental effects, such as contamination of the water and air [7], urban heat island effects, and the loss of agricultural land [8, 9]. Measurement and Identification of the urban growth trends and rate of sprawl can not only help to perform effective urban planning practices but also provide proper infrastructure [10]. Earth observation datasets by Remote sensing (RS) and GIS tools have been used extensively in sprawl and pattern-related studies of urbanization around the world [9, 11–13]. Urban growth studies provide very important information for planning purposes which is reported in many developed countries [6, 14–16]. Thus, effective urban planning is necessary to effectively manage the urbanization process and its negative consequences [17]. Monitoring and planning a city's growth is therefore essential for a city's proper development. A variety of disciplines, including geography, planning, and the social sciences, have studied this kind of phenomenon. Estimating urban growth can be a challenging process using traditional surveying and mapping techniques. Numerous urban growth studies have used statistical techniques and GIS [5, 9, 10, 18]. In addition to RS and GIS techniques, the Shannon entropy model has also been carried out to provide useful information. For a city to continue to flourish, urban growth management is essential and needs to be taken into consideration for urban sustainability. In urban, sub-non-rural, and agricultural research, the recognition of the non-rural area is critical [19, 20]. Gandhinagar features a Central Business District (CBD) and a Special Economic Zone (SEZ). These two zones have a greater impact on Gandhinagar's urbanization. Monitoring and evaluating a developing city like Gandhinagar have proved to be successful when using RS and GIS, and Shannon's entropy model together. The straightforward growth sign that can be monitored on the ground and examined in real-time data is the proportion of land covered by impermeable surfaces like concrete and asphalt. This is based on the hypothesis that developed, or built-up, regions have a larger proportion of impermeable surfaces than less developed ones. Additionally, the population of the area has an impact on the scattered area. This represents a quantifiable portion of the region's entire built-up area population. A similar technique is used to map sub-urban growth, and then contrast the resulting rural and urban centers. To better understand the behavior of such phenomena, sprawl is defined as an expansion in the built-up land along rural and urban peripheries. The built-up extent may be used as a good valid metric in urban sprawl predictions, according to earlier studies from around the globe.

Earlier studies investigated and applied geospatial methods to deliver a valuable contribution to the sector of urban energy [20]. Understanding the process and patterns of urbanization will be helpful for the current and future development scenarios of a region. This will also further be beneficial for different problem-solving purposes, i.e., to know the availability and potential of solar energy resources [21, 22] for further employment of eco-friendly technology. In the present work, remote sensing images over the period 1990-2020 were applied to examine the urban sprawl pattern and its quantity in Gandhinagar City, India. In particular, the present work is aimed at examining the urban land use changes and dynamics. This type of work is of great significance to policymakers, land managers, and local government departments at the regional level.

10.2 Study Area

With a total size of roughly 326 km^2 (126 sq m) and situated on the Sabarmati River's bank, Gandhinagar City may be found between latitude 23°12'56" N and longitude 72°38'12" E. Its typical altitude is 81 meters (266 feet). Gandhinagar has a tropical climatic condition with three main seasons: (1) summer, (2) monsoon and, (3) winter. The atmosphere is generally dry and hot outside of the rainy season. The weather is warm to extremely hot from March to May, when the highest temperature is in the range of 36 to 42 °C, and the lower temperature is in the range of 19 to 27 °C. During the winter months of December to February, it is comfortable during the day and rather chilly at night. The average maximum and minimum temperatures are about 29°C and 14°C respectively, and the environment is quite dry. The southwest rainy season conducts a moist or

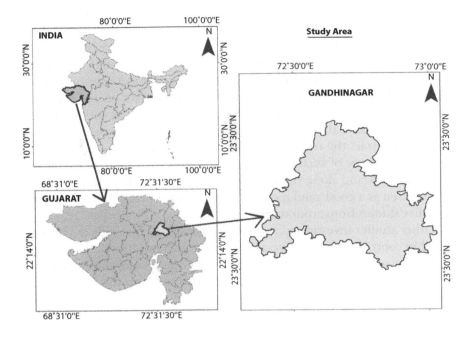

Figure 10.1 Location map for the present study.

cloudy climate from June through September. The range of annual rainfall is nearly 803.4 mm (31.63 in). The location map for the present study is presented in Figure 10.1.

10.3 Materials and Methodology

10.3.1 Satellite Data Used

In this study, multi-temporal satellite images of the Landsat series for the years 1990, 2000, 2010, and 2020 were acquired from the USGS website. The data for the population is collected from the Gandhinagar district population Census for 1991-2001, 2001-2011, and 2011-2021.

10.3.2 Pre-Processing of Satellite Data

All the collected satellite data (images) were atmospherically corrected and reprojected to the UTM coordinate system with the WGS-84 datum. For all the datasets, a relative band combination is chosen to generate False

Color Composition to identify and differentiate the various landscape features in the study region. All the photos from the research region were classified after applying the supervised classification technique. Figure 10.2 displays the specific methods used in the current investigation.

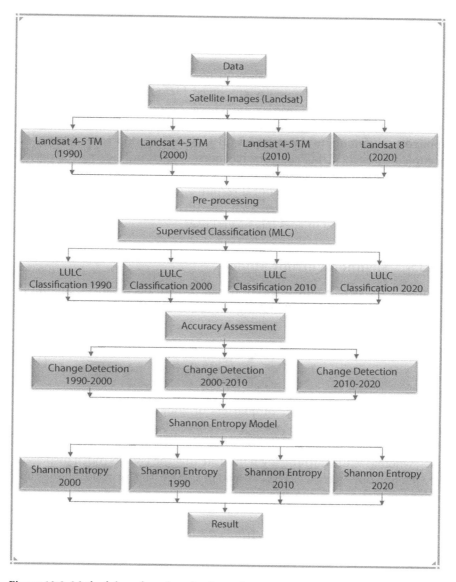

Figure 10.2 Methodology flow chart for the study.

10.3.3 Accuracy Assessment

It is required to compare the classified results or change the detection map to assess its accuracy and validation. The accuracy assessment was done by comparing the present study results with reference data. The kappa coefficient, user's and producer's accuracies, and total accuracy are all included in the accuracy evaluation.

10.3.4 LULC Change Detection

The change detection process aims to figure out the transitions among LULC categories. The frequently employed LULC change detection procedure consists of image overlay and classification differentiations of the statistics of LULC categories. In this study, supervised classification is applied to differentiate the statistics on LULC. A comparison of the areas occupied by each LULC category has been carried out during the period under investigation. Following that, each land cover feature's change in direction was identified. In natural resource management, environmental monitoring, and protection research, the identification of LULC changes using remote sensing images has several uses.

10.3.5 Shannon Entropy Model

To detect the extent and shape of urban growth occurrences, Shannon's entropy was estimated while taking the account urban expansion in various wards.

$$H_n = \sum P_i \log n (P_i) \quad (10.1)$$

Where n = total number of quarters or zones, Pi = proportion of the variable in the ith quarter or zone, and between 0 and log n is the ranges of entropy values. The value 0 denotes that the administration is very compact, whereas values around log n indicate that the issue is much dispersed. Greater entropy levels reveal the spread of sprawl. The formula for Shannon Index calculation is given as:

$$H' = -\sum [(n_1/N) * \ln(n_1/N)] \quad (10.2)$$

10.4 Results and Discussion

10.4.1 LULC Maps

A supervised classifier was applied in this work to obtain LULC maps for 1990, 2000, 2010, and 2020, respectively. The spatial distribution of different classes for three different years along with quantitative information, are shown in Figure 10.3 and Table 10.1, respectively. Furthermore, the urban growth during the periods 1990-2000, 2000-2010, and 2010-2020 are calculated and given in Table 10.2.

10.4.2 Accuracy Assessment

After classifying multi-temporal satellite images, the obtained accuracies indicate the reliability and usability of classified products. A dataset's two

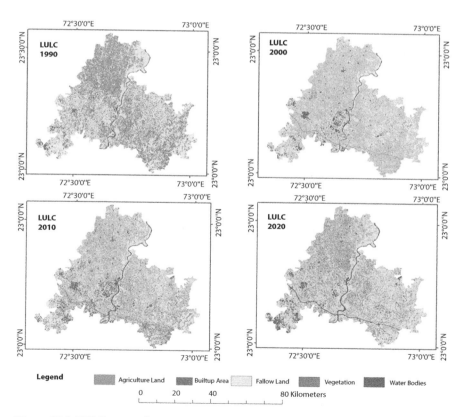

Figure 10.3 LULC maps of 1990, 2000, 20210, and 2020.

Table 10.1 Area distribution of LULC for 1990, 2000, 2010, and 2020.

Year	1990	2000	2010	2020
LULC	Area (km^2)	Area (km^2)	Area (km^2)	Area (km^2)
Agriculture Land	20.07	36.08	66.82	87.82
Built-up Land	60.58	86.45	125.04	148.18
Fallow Land	107.84	94.83	90.03	70.55
Vegetation Land	70.40	56.63	16.18	12.42
Water Bodies	67.11	52.01	27.93	7.03
Total	326	326	326	326

Table 10.2 Increase in the urban area (km^2) of Gandhinagar City, Gujarat.

Statistics for urban growth		
Year	Built-up area (km^2)	Increase in the built-up area (%)
1990	60.58	–
2000	86.45	42.70%
2010	125.04	44.64%
2020	148.18	18.51%

sets of categorizations are compared using the Kappa statistic to see how well they agree. In this study, several GCPs are used for assessing the classified results of years 1990, 2000, 2010, and 2020, respectively, using validation datasets from Google Earth images. The accuracy was performed separately for different years from 1990 to 2020.

The findings revealed that the producer's and user's accuracy varied from 75% to 100% and 75% to 95% individually. The overall accuracy between 1990 and 2020 was found to be 80%. Kappa statistics estimated the classification coefficient for 1990 and 2020 using the error matrix, and the results were 75% and 80%, respectively. Therefore, the categorization is in an excellent range according to the classification scale provided. Overall, it seems from the data that the accuracy was reasonable, with 87% and 86%

for the years 2000 and 2010, respectively. This outcome was superior to the 1990 categorized picture and met the minimal accuracy requirement for land cover change. The accuracy of the producer and user in this study spans from 78.94 to 93.33% and 70 to 100% in 2000, respectively, whereas it ranges from 80 to 93.75% and 75 to 93.75% in 2010. In the years 2000 and 2010, the classification's Kappa coefficient was 83.75% and 82.50%, respectively. As a result, according to the grading scale provided, the years 2000 and 2010 fall into the very good range and outstanding range, respectively.

10.4.3 LULC Change Detection

The LULC change detection based on the classified maps is performed to investigate the urban growth dynamics. The percentage of each LULC type is derived from the classified results for each year separately. Agriculture land, built-up land, fallow land, vegetation land, and water bodies changed significantly into other classes according to the change detection analysis. Moreover, the agricultural land, fallow land, and water bodies changed slightly into built-up land.

In 1990-2000, change detection shows the percentage of each area. The change detection of 1990-2000 shows high changes in the built-up land. There is a gradual conversion of water bodies and fallow land into built-up land. Vegetation land is slowly disappearing due to vegetation land taking place on agricultural land. The built-up part majorly covers the water bodies. It means the urban area is increasing near water bodies. From 2000-2010 change detection shows the percentage of each area. LULC Change detection of 2000-2010 shows the high changes in fallow land, it is gradually converting the water bodies and vegetation land into fallow land class. Vegetation land is slowly disappearing due to vegetation land taking place by agricultural land or fallow land. The built-up area majorly covers the water bodies. It means the urban area is increasing near water bodies. From 2010-2020 change detection shows the percentage of each area. LULC change detection of 2010-2020 shows the high changes in the built-up area; it is gradually converting the water bodies and fallow land in the built-up (urban) area. Vegetation land is slowly vanishing due to vegetation land changes to agricultural areas. The built-up area majorly covers the water bodies. It means the urban area is increasing near water bodies. The findings of change detection are presented in Figure 10.4.

The build-up area increased from 1990 to 2020 and the expansion of Gandhinagar City has grown from 60.58 km^2 to 148.18 km^2 (Table 10.1). The statistics of urban growth are given in Table 10.2. The statistics presented in Table 10.1 and Table 10.2 indicate that the number of

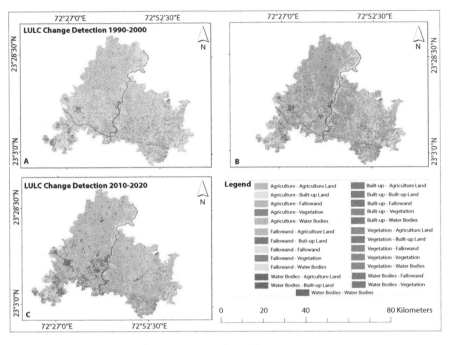

Figure 10.4 LULC change detection maps of Gandhinagar City, Gujarat.

developmental activities in Gandhinagar has outshined the growth rate of the population as given in Table 10.3. In the past last three decades, per capita consumption of land has grown quickly as the land is occupied for urbanization at a faster rate. All the lands for development initiatives such as residential, industrial, commercial, and educational are considered for inclusion in the per capita land consumption.

Between 2000 and 2010, both built-up areas and populations experienced the highest annual growth rates. Gandhinagar City experienced rapid physical growth during this time, which meant the highest level of land consumption per capita. As for the built-up areas, their growth rate slowed in the following period of 2010-2020. In this way, it can be explained why development rates have again risen. Both are now growing at a very rapid rate of about 4.28%. Figure 10.5 shows the urban development study using Shannon's entropy model.

Table 10.3 Population data of the study area.

S. no.	Years	Population data 1991-2021			Urban area (km²)	Rural area (km²)	References
		Total	Female	Male			
1	1991	408,992	195,563	213,429	167,219	241,773	Census 1991
2	2001	1,334,455	636,456	697,999	600,627	791,126	Census 2001
3	2011	1,391,753	667,889	723,864	376.59	1,763.41	Census 2011
4	2021	1,467,267	-		-	-	(estimates as per Aadhar uidai.gov. in Dec 2020 data)

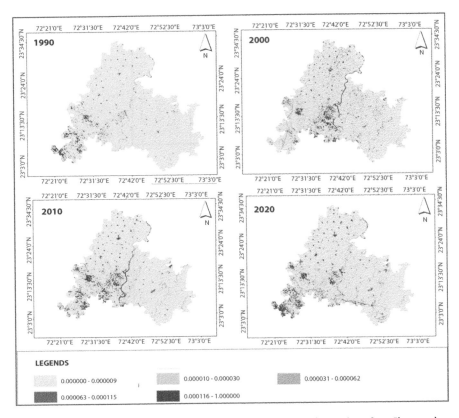

Figure 10.5 Urban growth for Gandhinagar between 1990 and 2020 based on Shannon's entropy model.

10.5 Conclusion

In this work, the change detection analysis based on multi-temporal remote sensing images and GIS provided vital information to comprehend the land use patterns useful for town planners and administrators. This region may thus be prepared for sustainable land management and city planning. Gandhinagar City has grown quickly during the last few decades. Remote sensing and GIS techniques integrated with the entropy approach effectively captured the spatial dynamics of urban growth. Greater Shannon entropy (greater than 1.6524) reflects the incidence and geographical allocation of the urban growth variables between 1990 and 2020. It is clear from this study that built-up areas around the city centre have been increasing continuously, with an outward expansion. The area of developed land has grown over time. The southeastern part of Gandhinagar has the highest percentage of increase in built-up land, while the southwestern part has the lowest. The present study results can also be used for sustainable practices, i.e., solar energy expansion. The southeastern talukas in Gandhinagar have a higher entropy value, indicating sprawl. It is also recognized that Gandhinagar town was developing in both a diverging and a straight path along the important roadways.

Acknowledgements

The authors are thankful to USGS for providing multi-temporal Landsat images for carrying out the present study.

References

1. A. Kundu, "Trends and patterns of urbanization and their economic implications", *India Infrastructure Report*, 29, 27-41, 2006.
2. B. N. Haack and A. Rafter, "Urban growth analysis and modelling in the Kathmandu valley, Nepal," *Habitat Int.*, vol. 30, no 4, pp. 1056–1065, 2006.
3. M. Punia and L. Singh, "Entropy approach for assessment of urban growth: a case study of Jaipur, India," *J. Indian Soc. Remote Sens.*, vol. 40, no. 2, pp. 231-244. 2012.
4. F. Yuan, K. E. Sawaya, B. C. Loeffelholz, M. E. Bauer, "Land cover classification and change analysis of the Twin Cities (Minnesota) metropolitan area by multitemporal Landsat remote sensing," *Remote Sens. Environ.*, vol. 98, no. 2-3, pp. 317-328, 2005.
5. V. N. Mishra and P. K. Rai, "A remote sensing aided multi-layer perceptron-Markov chain analysis for land use and land cover change prediction in Patna district (Bihar), India." *Arab. J. Geosci.*, vol. 9, no. 4, pp. 1-18, 2016.
6. A. Arora, M. Pandey, V. N. Mishra, R. Kumar, P.K. Rai, R. Costache, M. Punia, L. Di, "Comparative evaluation of geospatial scenario-based land change simulation models using landscape metrics," *Ecol. Indic.*, vol. 128, pp. 107810.
7. Y. Deng and S. Srinivasan, "Urban land use chan Cge and regional access: a case study in Beijing, China," *Habitat Int.*, vol. 51, pp. 103–113.
8. B. Huang, C. Xie, R. Tay, B. Wu, "Land-use-change modeling using unbalanced support-vector machines," *Environ. Plann B: Plann Des*, vol. 36, no. 3, pp. 398-416, 2009.
9. V.N. Mishra, P.K. Rai, R. Prasad, M. Punia, M. M. Nistor, "Prediction of spatio-temporal land use/land cover dynamics in rapidly developing Varanasi

district of Uttar Pradesh, India using Geospatial approach: a comparison of hybrid models," *Appl. Geomat.*, vol. 10, no. 3, pp. 257-276.
10. H. S. Sudhira, T. V. Ramachandra, K. S. Jagadish, "Urban sprawl: metrics, dynamics and modelling using GIS," *Int. J. Appl. Earth Obs. Geoinf.*, vol. 5, no. 1, pp. 29-39, 2004.
11. J. Cheng, I. Masser, "Urban growth pattern modelling: a case study of Wuhan City, PR China," *Landsc. Urban Plan.*, vol. 62, no. 4, pp. 199-217, 2003.
12. A. Chabaeva, D. L. Civco, S. Prisloe, "Development of a population density and land use-based regression model to calculate the amount of imperviousness," Doctoral dissertation, University of Connecticut, 2004.
13. V. N. Mishra, R. Prasad, P. Kumar, D. K. Gupta, S. Agarwal, A. Gangwal, "Assessment of spatio-temporal changes in land use/land cover over a decade (2000-2014) using earth observation datasets: a case study of Varanasi district, India," *Iran. J. Sci. Technol-Trans. Civ. Eng.*, vol. 43, pp. 383-401.
14. K. B. Barnes, J.M. Morgan, M.C. Roberge, S. Lowe, "Sprawl development: its patterns, consequences, and measurement." Towson University, Towson, 1, p. 24, 2001.
15. X. Yang, Z. Liu, "Use of satellite-derived landscape imperviousness index to characterize urban spatial growth," *Comput. Environ. Urban Syst.*, vol. 29, no. 5, pp. 524-540, 2005.
16. G. Li, Q. Weng, "Using Landsat ETM+ imagery to measure population density in Indianapolis, Indiana, USA," *Photogramm. Eng. Remote Sensing*, vol. 71, no. 8, pp. 947-958, 2005.
17. A. Soffianian, M. A. Nadoushan, L. Yaghmaei, S. Falahatkar, "Mapping and analyzing urban expansion using remotely sensed imagery in Isfahan, Iran," *World Appl. Sci. J.*, vol. 9, no. 12, pp. 1370-1378, 2010.
18. M. K. Jat, P. K. Garg, D. Khare, "Monitoring and modelling of urban sprawl using remote sensing and GIS techniques," *Int. J. Appl. Earth Obs. Geoinf.*, vol. 10, no. 1, pp. 26-43, 2008.
19. A.G.O. Yeh, X. Li, "Measurement and monitoring of urban sprawl in a rapidly growing region using entropy." *Photogramm. Eng. Remote Sensing*, vol. 67, no. 1, pp. 83-90, 2001.
20. H. S. Sudhira, T. V. Ramachandra, K. S. Raj, K. S. Jagadish, "Urban growth analysis using spatial and temporal data," *J. Indian Soc. Remote Sens.*, vol. 31, no. 4, pp. 299-311, 2003.
21. D. Kumar, "Mapping solar energy potential of southern India through geospatial technology," *Geocarto Int.*, vol. 34, no. 13, pp. 1477-1495, 2019.
22. D. Kumar, "Satellite-based solar energy potential analysis for southern states of India." *Energy Rep.*, vol. 6, pp. 1487-1500, 2020.

11
Sustainable Natural Spaces for Microclimate Mitigation to Meet Future Urban Energy Challenges

Richa Manocha[1*] and Deepak Kumar[1,2†]

[1]*Amity School of Business, Amity University Uttar Pradesh (AUUP), Gautam Buddha Nagar, Uttar Pradesh, India*
[2]*Amity Institute of Geoinformatics & Remote Sensing (AIGIRS), Amity University Uttar Pradesh (AUUP), Gautam Buddha Nagar, Uttar Pradesh, India*

Abstract

Streets, neighborhoods, and city planning affect the urban climate. Local climate change affects humans, energy, the built environment, and cities. In recent decades, urban climate science has evolved enough to be relevant in design and planning. Urban climate change's climatic, well-being, and carbon implications are summarized, along with effective mitigation techniques and their benefits. Map heat susceptibility and enhance heat resilience. This briefing note explains the difficulty of adjusting to the urban climate anomaly while planning new districts or infilling old ones. Connecting with others and the natural environment in metropolitan places is crucial to human well-being. Due to urbanites' hurried lifestyles and shrinking space, the divide between humans and nature is expanding. This research examines the literature on urban areas to draw links between humans and the environment, as well as the distance between them and strategies to bridge it. Urban greening can reestablish this link and reduce carbon footprint and energy use. Some say nature programs for impressionable children can help. These programs connect kids to nature. Existing structures are subject to climate change and create control issues. Urban regeneration and urban planning continue to prioritize tools that stimulate expansion within pre-existing urban districts, despite opposing demands.

---

*Corresponding author: richa.manocha1@gmail.com; Orcid: 0000-0001-7032-122X
†Corresponding author: deepakdeo2003@gmail.com; Orcid: 0000-0003-4487-7755

Deepak Kumar (ed.) Urban Energy Systems: Modeling and Simulation for Smart Cities, (171–192) © 2023 Scrivener Publishing LLC

Keywords: Green urban spaces, urban heat island (UHI), climate change, urban development, nature connections, urban canopy

11.1 Introduction

The climate of an urban region is significantly impacted by human activities. The behavior of the atmosphere over an extended period in a particular region, as characterized by elements such as temperature, pressure, wind, precipitation, cloud cover, and humidity, among other factors, is referred to as the climate [1–3]. In comparison to the regions that surround it, an urban area is characterized by a greater concentration of man-made buildings and other structures. In this discussion, we will investigate how human activities, like pollution, the color of buildings, the people who live there, factories, and so on, might alter the weather patterns that prevail in an urban setting [4, 5].

Both adaptation and mitigation are important parts of the fight against climate change and its effects on society and the environment. Mitigation aims to gradually cut emissions of climate-changing gases that are responsible for global warming [6, 7]. Adaptation seeks to reduce vulnerability of environmental, social, and economic systems while also increasing their capacity for climate resilience [8, 9]. There are several different ways to adapt to and mitigate the effects of climate change, but none of these approaches are sufficient on their own [10, 11]. The effectiveness of the implementation is dependent on policies and cooperation at all stages, and it can be improved by adopting integrated responses that link adaptation and mitigation with other social goals [12–14].

Human beings are socially oriented. They require good interpersonal interactions to achieve mental wellness [15]. They need to interact with other human beings in society as well as with nature around them. To be able to experience a purpose in one's life, it is important to feel as though one belongs to and is related to others. To feel like one belongs, one must experience relationships with other individuals, with society, and with nature [16–19]. However, in general, city life gives very little opportunity for such interaction. This is due to busy schedules, smaller families, physical distances, less access to open environments, etc. [20–22]. There is a scarcity of natural environments in urban spaces, resulting in a widespread concern that individuals are becoming more isolated because of the extinction of experiences from the natural world in both quality and quantity [23–25]. Recently, biophilia studies have emphasized the perspective of environmentally friendly behavior and the connection of human beings to nature. The concept emphasizes the fact that human beings' basic behavior is to relate to other forms of nature. The production of food is one way in which this connection can be established. An absence of this connection will result in an absence of knowledge of the food system and a disconnect with nature [26–28]. Citizens' knowledge of food is dwindling in urban areas due to shrinking space. The production of food is, in fact, a component of "nature," and a distance is said to exist between people and their environment in today's urban spaces [13, 29, 30]. Therefore, if a person loses their connection to nature, it may affect their relationship with food as well as the information they have about it. This paper explores the disconnect between human beings and their basic tendency to connect to nature. It also discusses how this disconnect is leading to less understanding of the importance of energy conservation through natural practices [14, 31, 32]. The study explores how reconnecting children at early levels through urban greening initiatives can result in bridging the divide thus created [33–35]. There are variety of planning scales that can be utilized to best effect to enhance urban microclimate, decrease heat islands, and increase water retention [36–39]. The several modeling approaches make it possible to estimate city-level energy demand and supply. Bottom-up methods are detailed and may model several technical choices; their primary goal is to identify the energy consumption value added by each end user [40–43]. The bottom-up strategy primarily makes use of statistical and engineering modeling techniques that are data-driven [44, 45]. Samples of customers' energy bills are used as inputs to statistical models that include factors like building age, shape, and occupant demographics to predict how much energy a structure would consume [46–48].

Large open spaces and forested regions, in addition to cold-air zones and cold-air snowboards, are crucial for urban areas [5, 49, 50]. Connected parks and green spaces, the preservation of rivers and open waterbodies, the building of artificial water surfaces, and large-scale retention/detention zones are crucial for heat-exposed locations [51, 52]. Decommissioning and unsealing impermeable areas are regarded as consistent ways to avoid hot spots in metropolitan regions [53–55]. It is strongly advised that parking lots, roadways, public locations, and buildings all have some form of shading. As a bonus, residential areas become more appealing when pocket parks and inside courtyards are landscaped. Water playgrounds, fountains, and medium-scale retention/detention basins are just some examples of the types of water surfaces seen in public spaces that can be designed to provide additional cooling. Green space irrigation should be integrated into the ecological planning of all water features [56, 57]. Greening the building's exterior and rooftop, as well as bolstering protection against the summer heat load (by measures like shade structures), are considered crucial components of cooling the building down. A second strategy involves making use of surfaces that reflect more light [57–59].

Pressure changes can also be caused by urban heat islands. A localized area of low pressure is created when warm air rises over a metropolitan area and absorbs heat from its surroundings. The addition to temperature and humidity; pressure may significantly alter the weather in a city. Winds can become extremely eddying when there is a large pressure difference between the windward and leeward sides of a building. When air moves from a high-pressure to a low-pressure area, the pressure difference between the two is the primary factor at play. The winds, in effect, are working to smooth out pressure variations around the planet. Due to high pressure, gentle country breezes blow from the chilly countryside to the warmer cities. As warm air rises, it generates a low-pressure area below it. Since winds tend to travel from high pressure to low pressure; this implies that pollution from the suburbs may be blown into the city [60, 61]. High pressure and calm conditions are ideal for observing contrasts and similarities between rural and urban locations. Undoubtedly, the velocity and the direction of winds in cities varies from those in the countryside.

11.2 Nature and Human Connection

Complexity is inherent in urban systems. Buildings, energy generation, transmission, and distribution networks, the local environment, and user participation in operational and strategic planning are all part of their energy modeling [62–64]. There has to be an examination of both intra- and inter-urban energy and resource movements [65–67]. To quantify the energy transition and to display results in ways that are useful to stakeholders, it is necessary to model and monitor changes over time because of socioeconomic and technological measures [68–70]. Urban infrastructure is notoriously complicated. Included in its energy modeling are structures, power plants, transmission and distribution systems, local ecosystems, and end-user input into operational and strategic planning. The transfer of energy and materials inside and between cities must be studied. Modeling and monitoring changes over time because of socioeconomic and technological interventions is important to quantify the energy transition and display results in ways that are usable by stakeholders [71].

Accurately assessing the real energy consumption in urban areas, in particular the consumption of the existing building stock is the first step in developing a long-term urban strategy. Most current energy consumption data is either aggregated at the national or international level, or it is very specific, such as for a single family. Energy modeling can help quantify the current state of actual consumption and the load profiles with higher time precision than is possible with metering alone, which is still expensive at the city scale. In addition, it enables the estimation of potential cost savings from implementing energy-efficiency measures and the prioritization of renovation projects [32, 72, 73].

As proposed by Nabhan *et al.* [69], the "biophilia theory" explains humans' urge to connect with nature. Wilson believes that we have an underlying urge to connect with all other species because we have developed in nature and are highly dependent on nature for our survival, therefore we have an inbuilt need to connect with all other life forms [74, 75]. The biophilia theory argues that individuals are naturally programmed and motivated to seek out and enjoy nature for their very existence. However, some people may hate being in nature owing to conditioning by environmental factors like peers, parents, or unpleasant personal experiences in nature. As the meaning and perception of nature are changing, it is difficult to expect everyone to be connected to nature in the same way. Similarly, a rural perspective on nature may differ from that of a city dweller, and even across various social and cultural subgroups within the same area. However, we can broadly define nature as per the various theorists, indoor plants, urban nature reserves, urban vegetation, agricultural land, and countryside natural vegetation like a forest, woodland, etc. [76–79]. Production of food in cities and outside is considered an important part of nature. Past positive experiences with nature help create feelings of affiliation, calmness, and peace which may lead to pro-environmental or protective behavior toward nature. The biophilia theory has been used to help explain the reasons why people do or do not want to protect the environment. It is critical to remember that just because individuals do not act in environmentally responsible ways does not imply that they are indifferent to the environment. However, a connection is established between people who relate to nature and the time they spend in the natural environment and the extent of support they provide towards pro-environmental behavior. Being with nature has been shown to improve physical health, mental health, psychological health, psychological wellness, life meaning, cognitive capacity, and social cohesiveness, as well as offer areas for physical and social activity [59, 80, 81].

Within the framework of urban planning tools, the selection of measures should be based on criteria of efficiency and local practicability. A model can be used to validate the measures and their effects, or the literature can be consulted for confirmation. The applicability of local measures, and consequently their transferability, is very context dependent. Respectful urban planning and climate-proofing measures are especially difficult to implement in historic structures. The use of unsealed pavement, the planting of shade trees, and the incorporation of water features into public spaces are all tried-and-true methods for reducing urban heat [82–84]. Both methods, however, run counter to the standards of monument protection and the original intent of classical and baroque plazas.

11.3 Urban Gardening

Many modern cities are rapidly losing food-growing areas and abilities, and we are already forgetting how to cultivate food. The important connection with nature that human beings establish by growing food is declining among urban citizens [85–87]. This has been studied by scholars from the fields of foodscapes and urban greening. However, the consequences for human health and welfare are frequently lacking. Food is important to our health and well-being, so it is seen as "the ultimate measure of our ability to care for ourselves and others." Urban greening is the practice of managing open spaces within cities and creating a liveable place surrounded by green spaces and recreation sites [88, 89]. According to studies, merely seeing a plant can cause people to feel relaxed, calm, and confident, which can help reduce blood pressure and muscular tension [90, 91]. Community gardens are found to be calming places and "raise nutrition and physical activity while promoting the importance of public health in enhancing the quality of life". The incorporation of urban greening within a city is a strategy for using "green spaces" to help improve public health [92–94]. Some examples of urban greening are community gardens, home gardens, rooftop gardens, balconies, and urban farms. We are witnessing new agricultural technologies, like vertical farming, rooftop farming, and floating hydroponic farms being constructed in urban spaces [30, 95, 96]. However, the economic viability and acceptability of such projects are concerns in the long run. In low-income nations, the practice of urban farming is done for local consumption, monetary advantages, and agricultural traditions, as well as a source of cheap food with diversity and quality. In contrast, urban greening in developed countries is undertaken for more personal and psychological reasons, such as a requirement for high-quality, fresh, organic food; prior exposure; health reasons; and environmental convictions [97, 98].

11.4 Urban Greening and Energy Benefits

One of the arguments for considering urban greening has been its possible environmental advantages, such as lower energy use across the food supply chain. Advocates of urban greening claim that reintroducing food production in cities results in several energy-related benefits [99–101]. Research majorly highlights energy saving due to reductions in transportation, wholesale and resale storage requirements, reduction of food waste along the distribution network and easier resource exploitation [47, 101–103]. Peri-urban greening has the potential to deliver less resource-intensive output by preserving higher-yielding prime agricultural land. Exploiting waste streams in more complex integrated operations (vertical farms, integrated greenhouses) might balance energy needs for delivering these inputs in conventional operations. Furthermore, if urban greening dispersed nature is backed by a similarly distributed energy infrastructure system, food/agriculture waste may be processed locally to create biogas for heat or power generation, reducing the urban greening energy footprint even further [13, 104, 105]. The reuse of both nutrients and water from home-treated wastewater in urban greening is frequently seen as a viable food-energy-water-health (FEW) solution. FEW-health is an approach that may involve the utilization of energy while employing water recycling and nutritional, recovery while at the same time offering access to nutritious meals [104–106]. The FEW approaches are key strategies for sustainable infrastructure in cities. It is an approach that is considered to promote successful project implementation. Water recycling in the case of urban greening results in benefits like nutrient recycling, reduced fertilizer usage, treatment of wastewater, accessibility of irrigation water, and reduced need for expensive refrigerated transport or storage facilities [5, 13, 49, 107–110].

11.5 Nurturing a Connection to Nature in Early Years

Indoor play, the use of gadgets, and changes in parenting attitudes are just a few of the major reasons for the drastic changes in childhood that have occurred in the last few decades. Busy and stressful work schedules and constant exposure to accidents cause parents to be too concerned to allow their children unrestricted play outside. The radius within which children may freely walk outside their houses has shrunk by 90% since the 1970s. Similarly, in a school context, a shortage of open areas, a rigorous curriculum, parental expectations, and fear of repercussions such as accidents are restricting the use of the outdoors as a "natural classroom." This has resulted in "nature deficiency" among children. This is also true for food since children are growing increasingly separated from the land and communities that generate their food. Students have very limited knowledge of the production process of the food that they eat and an understanding of farming operations. Adolescence is marked by a physical and psychosocial transition, a shift in cognitive processing, increased independence in decision-making, increased responsibilities and experimentation, and the development of personal identity. This results in increased tension and worry. Experiences like gardening have been proven to boost children's empathy for the natural world as well as their perceptual abilities, self-esteem, self-acceptance, and self-efficacy, along with their capability of coping better with hardship. Nature mitigates the detrimental impacts of stressful conditions by offering psychological repair. Simple treatments, such as encouraging individuals to recognize good things in and around nature, have been shown to improve people's connection to nature. Such programs have proven that connecting with the environment may help people build resilience to deal with the obstacles of life while also promoting exercise, social contact, and a sense of purpose. Programs that expose children to the natural environment have been demonstrated to boost their emotional affinity for nature, ecological ideas, and readiness to exhibit ecological behavior. Environmental programs and direct nature exposure have been shown to have a greater impact on environmental behavior than environmental education alone. Childhood is a time when individuals are most flexible, as their beliefs and habits are yet to be completely formed, so it is easier to mold their habits, attitudes, preferences, and behavior. Thus, it is a good time to establish a positive attitude towards nature and food production. Knowledge of food, nutrition and health, agriculture, and skills of planning, preparing, and eating food are all important components of food literacy, which is integral to an individual's empowerment in the food system. Food literacy programs in schools are important to provide food and nutritional knowledge and skills among individuals. Food literacy should begin at an elementary level and extend until high school as young people are cognitively, emotionally, and intellectually capable of analysis. Research output shows that adolescents are well-informed about good health and nutrition habits, but this information is not translated into healthy dietary behavior. Adolescents look to their parents and teachers to encourage, support and enable them to be engaged in more healthy behaviours. A few student groups also hold their parents entirely responsible for meal preparation and blame them for their poor eating habits.

In the recent decade, the literature has been expanded by a complex body of information on Geospatial technology in urban climate change mitigation and adaptation. Many studies provide actual evidence for designing geographic information to reduce a city's climate change risk. Studies have highlighted the importance of geographic information systems in mitigating climate change and balancing urban carbon emissions (particularly with green roofs and green facades); thermal comfort due to the cooling effect of green roofs in different types of buildings and seasons; and other positive effects. Socially vulnerable individuals can benefit from social and psychological support.

Because of their lengthy history of evolution in the environment, human beings have learned to adapt easily to a variety of conditions. There must be parks, plazas, and other green areas in every city design so that residents have access to nature and the city's numerous benefits. In contrast, natural environmental resources have been harmed by uncontrolled development and the transformation of green spaces into grey buildings. Numerous studies have demonstrated the positive effects of proximity to green space in metropolitan areas on human well-being, and it is now time to determine the precise function these areas provide. Population shifts, variations in building style, urban sprawl, topography, and other factors can all have a bearing on the types of green spaces that are needed in cities.

Planning decisions made at the block, neighborhood, and even city level can have both intended and unforeseen consequences for the local environment. This is what we mean when we talk about the urban climate. This kind of local climate change contributes to the overall alteration of the global climate, where it not only interacts with but also exacerbates the effects of climate change on humans, the energy sector, the built environment, and metropolitan areas. Even though the study of urban climate is still a relatively new subject of research, there has been sufficient growth in knowledge about it in recent decades for it to be of practical value to decision-making in the areas of design and planning. The effects of urban climate change on climate, well-being, and carbon are summarized, along with the most effective methods for mitigating its effects and the relative advantages of those practices. Mapping areas that are susceptible to heat and developing strategies to better withstand its effects are two essential action points. Whether it is the process of developing new districts or in the process of infilling existing ones, it is anticipated that this briefing note will enhance understanding of the vast variety of challenges that are involved in reacting to the urban climate anomaly.

The effects of climate change combined with the ongoing trends in urbanization make it imperative that cities become more resilient. Heat waves are a common problem in urban settings, and as a result, cities typically have to use a significant amount of energy to keep their structures cool.

Floods and the damage they do are made worse by impervious surfaces because urban runoff is swiftly released into receiving water bodies, where it has the potential to upset aquatic ecosystems and is frequently a substantial source of pollution. This makes floods more likely to occur.

11.6 Conclusion

Because of human activity, the temperature in an urban microclimate is greater than the temperature that is typically expected in the surrounding areas and what it should be. It's been stated that cities are like little heat islands. This is especially obvious when conditions are quiet because the city center will be several degrees warmer than the countryside at that time. The human influence on urban microclimates causes them to be warmer than their rural counterparts. The term "urban heat island" refers to the fact that, even on a mild day, temperatures tend to rise in the heart of a city and fall outwards into the suburbs and rural areas. This recurring trend can be attributed to a few distinct causes. As a result, heat is retained in the air, making cities unattractive for living in. The dark colour and high thermal capacity of common road surfaces like tarmac and concrete mean they absorb a lot of heat. This heat is stored during the day and gradually released during the evening, causing an increase in temperature. The city's manufacturing and increased car use release additional heat, contributing to pollution that leads to smog and a pollution dome. This pollution dome lets in short-wave insolation but traps outgoing terrestrial radiation with a longer wavelength, boosting the amount of heat obtained. Health and well-being depend on natural relationships. These relationships must be cultivated in early childhood to establish a deep bond with nature. This will lead to health benefits like blood pressure regulation, psychological benefits like growing resilience, and food security through an understanding of food growth. Healthy eating habits involve more than knowing about nutritious meals; they also involve developing a positive relationship with food. Combining environmental and food-related activities can increase this link. These activities build informal social support, trust, and cohesiveness, which alleviate food insecurity. These programs provide new markets for farmers and help build a resilient food system.

11.7 Future Implication

Agriculture in urban areas has repercussions beyond the kitchen. More study is needed to develop efficient preventative interventions using a transdisciplinary approach, as industrialized cultures spread to urban areas in underdeveloped countries. Today, it is crucial to comprehend the actions of all the involved parties in a dynamic and intricate system. It will be the academic responsibility of future planners to offer superior personnel training to postgraduate, doctoral, and postdoctoral students, in addition to industry professionals. Students and industry professionals alike will be instructed to study the urban microclimate in addition to the science behind UHI. In addition to this, they will be instructed in the proper use of the devices that are used for detecting radiation and will be tasked with collecting data that will be fed into the numerical models that are used to simulate the urban microclimate. The research students need to be encouraged to participate in new studies that are aimed at finding solutions to the issues that arise from urban climate. Currently, urban people are disconnected from the processes involved in producing food and the effects on human health and the environment. There is a pressing need to better understand the links since this poses a serious threat to the future of creating livable cities.

References

1. Baniya, Binod, Kua-anan Techato, Sharvan Kumar Ghimire, and Gyan Chhipi-shrestha. 2018. "A Review of Green Roofs to Mitigate Urban Heat Island and Kathmandu Valley in Nepal." *Applied Ecology and Environmental Sciences* 6(4):137–52. doi: 10.12691/aees-6-4-5.
2. Baqa, M. F., F. Chen, L. Lu, S. Qureshi, A. Tariq, S. Wang, L. Jing, S. Hamza, and Q. Li. 2021. "Monitoring and Modeling the Patterns and Trends of Urban Growth Using Urban Sprawl Matrix and CA-Markov Model: A Case Study of Karachi, Pakistan." *Land* 10(7). doi: 10.3390/land10070700.
3. Basova, S., A. Sopirova, and E. Putrova. 2020. "Development of a New Urban Country on the Danube Arms near Bratislava." in *IOP Conference Series: Materials Science and Engineering*. Vol. 960, edited by D. M. Yilmaz I. Marschalko M. IOP Publishing Ltd.
4. Bayulken, B., D. Huisingh, and P. M. J. Fisher. 2021. "How Are Nature Based Solutions Helping in the Greening of Cities in the Context of Crises Such as Climate Change and Pandemics? A Comprehensive Review." *Journal of Cleaner Production* 288. doi: 10.1016/j.jclepro.2020.125569.

5. Bendito, América. 2020. "Grounding Urban Resilience through Transdisciplinary Risk Mapping." *Urban Transformations* 2(1):1–11. doi: 10.1186/s42854-019-0005-3.
6. Bolay, J. C., S. Cartoux, A. Cunha, T. T. N. Du, and M. Bassand. 1997. "Sustainable Development and Urban Growth: Precarious Habitat and Water Management in Ho Chi Minh City, Vietnam." *Habitat International* 21(2):185–97. doi: 10.1016/S0197-3975(97)89956-0.
7. Boza Valle, J. A., E. Y. Mendoza Vargas, H. E. Escobar Terán, and O. F. Moncayo Carreño. 2020. "Infrastructure of the Hotel Entrepreneurship in the Quevedo Canton Regarding Its Demand [Infraestructura de Los Emprendimientos Hoteleros En El Cantón Quevedo Respecto a Su Demanda]." *Universidad y Sociedad* 12(1):337–42.
8. Bravet, P., A. MARCO, and V. B. MONTÈS. 2018. "Characterisation of the Vegetation of Urban Wastelands Situated within Transport Infrastructure Sites: The Case of the Huveaune Valley – Marseille." *Ecologia Mediterranea* 44(1):67–83. doi: 10.3406/ecmed.2018.2030.
9. Buckerfield, S. J., R. S. Quilliam, L. Bussiere, S. Waldron, L. A. Naylor, S. Li, and D. M. Oliver. 2020. "Chronic Urban Hotspots and Agricultural Drainage Drive Microbial Pollution of Karst Water Resources in Rural Developing Regions." *Science of the Total Environment* 744. doi: 10.1016/j.scitotenv.2020.140898.
10. Bullock, E. L., C. E. Woodcock, and P. Olofsson. 2020. "Monitoring Tropical Forest Degradation Using Spectral Unmixing and Landsat Time Series Analysis." *Remote Sensing of Environment* 238. doi: 10.1016/j.rse.2018.11.011.
11. Capolupo, A., C. Monterisi, C. Barletta, and E. Tarantino. 2021. "Google Earth Engine for Land Surface Albedo Estimation: Comparison among Different Algorithms." in *Proceedings of SPIE - International Society for Optical Engineering*. Vol. 11856, edited by M. A. Neale C.M.U. SPIE.
12. Capolupo, A., C. Monterisi, A. Sonnessa, G. Caporusso, and E. Tarantino. 2021. "Modeling Land Cover Impact on Albedo Changes in Google Earth Engine Environment" edited by M. S. G. C. B. I. T. D. A. B. O. R. A. M. T. E. T. C. M. Gervasi O. Murgante B. *Lecture Notes in Computer Science (Including Subseries Lecture Notes in Artificial Intelligence and Lecture Notes in Bioinformatics)* 12955 LNCS:89–101. doi: 10.1007/978-3-030-87007-2_7.
13. Chan, Isabelle Y. S., and Anita M. M. Liu, "Effects of Neighborhood Building Density, Height, Greenspace, and Cleanliness on Indoor Environment and Health of Building Occupants." *Build. Environ. 2018 Nov; 145*, pp. 213–22. doi: 10.1016/j.buildenv.2018.06.028.
14. de Caro, M., G. Crosta, P. Frattini, R. Castellanza, F. Tradigo, A. Mussi, and P. Cresci. 2019. "Blue-Green Infrastructures and Groundwater Flow for Future Development of Milano (Italy) [Infrastructures Bleu-Vert et Écoulement Des Eaux Souterraines Pour Le Développement Futur de Milan (Italie)]." in *17th European Conference on Soil Mechanics and Geotechnical Engineering, ECSMGE 2019 - Proceedings*. Vols. 2019-Septe, edited by E. S. B.

B. Sigursteinsson H. Erlingsson S. International Society for Soil Mechanics and Geotechnical Engineering.
15. Crainic, R., and R. Fechete. 2020. "Advanced Monitoring of a Laboratory Scale Modular Green Roof Model." in *AIP Conference Proceedings*. Vol. 2206, edited by T. I. Colnita A. American Institute of Physics Inc.
16. Crombé, L., and D. Blanchon. 2010. "Micro-Networks and the Reconquest of the City: A Case Study in Khartoum [Les (Micro)-Réseaux à La Reconquête de La Ville: Le Cas de Khartoum]." *Bulletin d'Association de Geographes Francais* 87(4):517–33. doi: 10.3406/bagf.2010.8195.
17. Cuca, B., and A. Agapiou. 2017. "Impact of Land Use Change to the Soil Erosion Estimation for Cultural Landscapes: Case Study of Paphos Disrict in Cyprus." Pp. 25–29 in *International Archives of the Photogrammetry, Remote Sensing and Spatial Information Sciences - ISPRS Archives*. Vol. 42.
18. Dall'O', G. 2020. "Renaturing Cities: Green and Blue Urban Spaces as Paradigms of Urban Planning." *Research for Development* 43–65. doi: 10.1007/978-3-030-41072-8_3.
19. Dover, J. W. 2018. *Introduction to Urban Sustainability Issues: Urban Ecosystem*. Elsevier Inc.
20. Engemann, K., J. C. Svenning, L. Arge, J. Brandt, C. Erikstrup, C. Geels, O. Hertel, P. B. Mortensen, O. Plana-Ripoll, C. Tsirogiannis, C. E. Sabel, T. Sigsgaard, and C. B. Pedersen. 2020. "Associations between Growing up in Natural Environments and Subsequent Psychiatric Disorders in Denmark." *Environmental Research* 188. doi: 10.1016/j.envres.2020.109788.
21. Facillity, Alaska Satellite. 2017. "Land Cover Change Detection with S-1TBX: Create an RGB from Multi-Temporal Sentinel-1 Data A) System Requirements." (September 2017).
22. Fargas D.C., Jr., G. A. M. Narciso, and A. C. Blanco. 2021. "Monitoring and Assessment of Agri-Urban Land Conversion Using Multi-Sensor Remote Sensing and GIS Techniques." Pp. 117–24 in *ISPRS Annals of the Photogrammetry, Remote Sensing and Spatial Information Sciences*. Vol. 5, edited by L. F. Y. M. Y. J. S. S. A. Z. H. L. X. O. B. S. U. H. E. S. M. Z. J. P. A. W. L. L. R. Y. M. D. K. A. O. A. H. M. F. F. S. Paparoditis N. Mallet C. Copernicus GmbH.
23. Farzaneh, N., C. A. Williamson, J. Gryak, and K. Najarian. 2021. "A Hierarchical Expert-Guided Machine Learning Framework for Clinical Decision Support Systems: An Application to Traumatic Brain Injury Prognostication." *NPJ Digital Medicine* 4(1). doi: 10.1038/s41746-021-00445-0.
24. Friedrich, C. 2008. "Selecting the Proper Components for a Green Roof Growing Media." Pp. 240–51 in *Low Impact Development: New and Continuing Applications - Proceedings of the 2nd National Low Impact Development Conference 2007*. Vol. 331. Wilmington, NC.
25. Garrett, J. K., M. P. White, L. R. Elliott, B. W. Wheeler, and L. E. Fleming. 2020. "Urban Nature and Physical Activity: Investigating Associations Using

Self-Reported and Accelerometer Data and the Role of Household Income." *Environmental Research* 190. doi: 10.1016/j.envres.2020.109899.
26. Ghasempour, F., A. Sekertekin, and S. H. Kutoglu. 2021. "Google Earth Engine Based Spatio-Temporal Analysis of Air Pollutants before and during the First Wave COVID-19 Outbreak over Turkey via Remote Sensing." *Journal of Cleaner Production* 319. doi: 10.1016/j.jclepro.2021.128599.
27. Gidlow, C., E. van Kempen, G. Smith, M. Triguero-Mas, H. Kruize, R. Gražulevičienė, N. Ellis, G. Hurst, D. Masterson, M. Cirach, M. van den Berg, W. Smart, A. Dėdelė, J. Maas, and M. J. Nieuwenhuijsen. 2018. "Development of the Natural Environment Scoring Tool (NEST)." *Urban Forestry and Urban Greening* 29:322–33. doi: 10.1016/j.ufug.2017.12.007.
28. Greca, P. L., L. Barbarossa, M. Ignaccolo, G. Inturri, and F. Martinico. 2011. "The Density Dilemma. A Proposal for Introducing Smart Growth Principles in a Sprawling Settlement within Catania Metropolitan Area." *Cities* 28(6):527–35. doi: 10.1016/j.cities.2011.06.009.
29. Guerrero Delgado, M. C., J. Sánchez Ramos, M. C. Pavón Moreno, J. A. Tenorio Ríos, and S. Álvarez Domínguez. 2020. "Experimental Analysis of Atmospheric Heat Sinks as Heat Dissipators." *Energy Conversion and Management* 207. doi: 10.1016/j.enconman.2020.112550.
30. Gunawardena, K. R., M. J. Wells, and T. Kershaw. 2017. "Utilising Green and Bluespace to Mitigate Urban Heat Island Intensity." *Science of the Total Environment* 584–585:1040–55. doi: 10.1016/j.scitotenv.2017.01.158.
31. Gupta, R., N. Budhiraja, S. Mago, and S. Mathur. 2021. *An IoT-Based Smart Parking Framework for Smart Cities*. Vol. 1174.
32. Haase, D., M. Wolff, and N. Schumacher. 2021. "Mapping Mental Barriers That Prevent the Use of Neighborhood Green Spaces." *Ecology and Society* 26(4). doi: 10.5751/ES-12675-260416.
33. Halder, Bijay, Jatisankar Bandyopadhyay, and Papiya Banik. 2021. "Monitoring the Effect of Urban Development on Urban Heat Island Based on Remote Sensing and Geo-Spatial Approach in Kolkata and Adjacent Areas, India." *Sustainable Cities and Society* 74(March):103186. doi: 10.1016/j.scs.2021.103186.
34. Han, J., Z. Zhang, J. Cao, Y. Luo, L. Zhang, Z. Li, and J. Zhang. 2020. "Prediction of Winter Wheat Yield Based on Multi-Source Data and Machine Learning in China." *Remote Sensing* 12(2). doi: 10.3390/rs12020236.
35. Hayat, Parvez. 2016. "Smart Cities: A Global Perspective." *India Quarterly* 72(2):177–91. doi: 10.1177/0974928416637930.
36. He, Y., T. Liu, T. Wang, X. Liang, H. Wei, Z. Zheng, and X. Xiao. 2021. "Case Analysis of Integrated Maintenance Technology for Multi-Dimensional Rapid Detection and Trenchless Reinforcement." in *E3S Web of Conferences*. Vol. 261, edited by M. S. Mostafa M.M.H. EDP Sciences.
37. Hsu, C. Y., and S. J. Ou. 2019. "A Study on Application of Ecological Engineering Methods to a River Pollution Remediation - Case Study of

Liuchuan River." in *IOP Conference Series: Earth and Environmental Science*. Vol. 291. Institute of Physics Publishing.
38. Hua, A. K., and O. W. Ping. 2018. "The Influence of Land-Use/Land-Cover Changes on Land Surface Temperature: A Case Study of Kuala Lumpur Metropolitan City." 51(1):1049–69. doi: 10.1080/22797254.2018.1542976.
39. Ile, U., and A. Ziemelniece. 2019. "Green-Blue Infrastructure in Multi-Storey Residential Area." in *IOP Conference Series: Materials Science and Engineering*. Vol. 603, edited by D. M. D. A. M. T. D. N. D. Yilmaz I. Marschalko M. Institute of Physics Publishing.
40. Jayaraman, V., D. Gowrisankar, and S. K. Srivastava. 2006. "India's EO Pyramid for Holistic Development." Pp. 2182–92 in *AIAA 57th International Astronautical Congress, IAC 2006*. Vol. 4. Valencia: American Institute of Aeronautics and Astronautics Inc.
41. Jena, L. K. 2021. "Does Workplace Spirituality Lead to Raising Employee Performance? The Role of Citizenship Behavior and Emotional Intelligence." *International Journal of Organizational Analysis*. doi: 10.1108/IJOA-06-2020-2279.
42. Keller, S. B., P. M. Ralston, and S. A. LeMay. 2020. "Quality Output, Workplace Environment, and Employee Retention: The Positive Influence of Emotionally Intelligent Supply Chain Managers." *Journal of Business Logistics* 41(4):337–55. doi: 10.1111/jbl.12258.
43. Klemeš, Jiří Jaromír, Petar Sabev Varbanov, and Donald Huisingh. 2012. "Recent Cleaner Production Advances in Process Monitoring and Optimisation." *Journal of Cleaner Production*. doi: 10.1016/j.jclepro.2012.04.026.
44. Klink, R. R., J. Q. Zhang, and G. A. Athaide. 2021. "Measuring Customer Experience Management and Its Impact on Financial Performance." *European Journal of Marketing* 55(3):840–67. doi: 10.1108/EJM-07-2019-0592.
45. Knapp, S., S. Schmauck, and A. Zehnsdorf. 2019. "Biodiversity Impact of Green Roofs and Constructed Wetlands as Progressive Eco-Technologies in Urban Areas." *Sustainability (Switzerland)* 11(20). doi: 10.3390/su11205846.
46. Koesling, M., N. P. Kvadsheim, J. Halfdanarson, J. Emblemsvåg, and C. Rebours. 2021. "Environmental Impacts of Protein-Production from Farmed Seaweed: Comparison of Possible Scenarios in Norway." *Journal of Cleaner Production* 307. doi: 10.1016/j.jclepro.2021.127301.
47. Krentowski, J., S. Mlonek, K. Ziminski, and A. Tofiluk. 2019. "Structural and Technological Aspects of the Historical Floors Replacement." in *IOP Conference Series: Materials Science and Engineering*. Vol. 471, edited by C. E. R. J. D. A.-M. M. D. M. S. A. Yilmaz I. Drusa M. Institute of Physics Publishing.
48. kullo, Emmanuel Daata, Eric Kwabena Forkuo, Ernest Biney, Emmanuel Harris, and Jonathan Arthur Quaye-Ballard. 2021. "The Impact of Land Use and Land Cover Changes on Socioeconomic Factors and Livelihood in the Atwima Nwabiagya District of the Ashanti Region, Ghana." *Environmental Challenges* 5(May):100226. doi: 10.1016/j.envc.2021.100226.

49. Kumar, Deepak. 2015. "Economic Assessment of Photovoltaic Energy Production Prospects in India." *Procedia Earth and Planetary Science* 00.
50. Kumar, Deepak, and Sulochana Shekhar. 2014. "Photovoltaic Energy Asessment Using Geospatial Technology." *Internation Journal of Scientific & Technology Research* 3(6):54–60.
51. Kumar, Deepak, and Sulochana Shekhar. 2015. "Statistical Analysis of Land Surface Temperature – Vegetation Indexes Relationship through Thermal Remote Sensing." *Ecotoxicology and Environmental Safety* (2):1–6. doi: 10.1016/j.ecoenv.2015.07.004.
52. Kumar, Deepak, and Sulochana Shekhar. 2016. "Linear Gradient Analysis of Kinetic Temperature through Geostatistical Approach." *Modeling Earth Systems and Environment* 2(3). doi: 10.1007/s40808-016-0198-3.
53. Kumari, N., O. Yetemen, A. Srivastava, J. F. Rodriguez, and P. M. Saco. 2019. "The Spatio-Temporal Ndvi Analysis for Two Different Australian Catchments." Pp. 958–64 in *23rd International Congress on Modelling and Simulation - Supporting Evidence-Based Decision Making: The Role of Modelling and Simulation, MODSIM 2019*, edited by E. S. Modelling and Simulation Society of Australia and New Zealand Inc. (MSSANZ).
54. Laskar, N., U. Singh, R. Kumar, and S. K. Meena. 2022. "Spring Water Quality and Assessment of Associated Health Risks around the Urban Tuirial Landfill Site in Aizawl, Mizoram, India." *Groundwater for Sustainable Development* 17. doi: 10.1016/j.gsd.2022.100726.
55. Lathrop, R., and J. Hasse. 2006. *Tracking New Jersey's Changing Landscape*. Vol. 9780813539. Rutgers University Press.
56. Li, J., L. Zhang, and H. Ye. 2021. "Spatiotemporal Change Analysis of Annual Average NDVI in Qinling Mountains as Ecological Security Barrier and Dividing Line between Geography and Climate." in *IOP Conference Series: Earth and Environmental Science*. Vol. 804. IOP Publishing Ltd.
57. Li, W., H. El-Askary, R. Thomas, S. P. Tiwari, K. P. Manikandan, T. Piechota, and D. Struppa. 2020. "An Assessment of the Hydrological Trends Using Synergistic Approaches of Remote Sensing and Model Evaluations over Global Arid and Semi-Arid Regions." *Remote Sensing* 12(23):1–28. doi: 10.3390/rs12233973.
58. Lin, B. B., M. H. Egerer, H. Liere, S. Jha, P. Bichier, and S. M. Philpott. 2018. "Local- and Landscape-Scale Land Cover Affects Microclimate and Water Use in Urban Gardens." *Science of the Total Environment* 610–611:570–75. doi: 10.1016/j.scitotenv.2017.08.091.
59. Lin, M. H., J. Hu, M. L. Tseng, A. S. F. Chiu, and C. Lin. 2016. "Sustainable Development in Technological and Vocational Higher Education: Balanced Scorecard Measures with Uncertainty." *Journal of Cleaner Production* 120:1–12. doi: 10.1016/j.jclepro.2015.12.054.
60. Lourdes, K. T., C. N. Gibbins, P. Hamel, R. Sanusi, B. Azhar, and A. M. Lechner. 2021. "A Review of Urban Ecosystem Services Research in Southeast Asia." *Land* 10(1):1–21. doi: 10.3390/land10010040.

61. Lundy, L., and R. Wade. 2011. "Integrating Sciences to Sustain Urban Ecosystem Services." *Progress in Physical Geography* 35(5):653–69. doi: 10.1177/0309133311422464.
62. Luo, H., and J. Wu. 2021. "Effects of Urban Growth on the Land Surface Temperature: A Case Study in Taiyuan, China." *Environment, Development and Sustainability* 23(7):10787–813. doi: 10.1007/s10668-020-01087-0.
63. Madurai Elavarasan, R., and R. Pugazhendhi. 2020. "Restructured Society and Environment: A Review on Potential Technological Strategies to Control the COVID-19 Pandemic." *Science of the Total Environment* 725. doi: 10.1016/j.scitotenv.2020.138858.
64. Manusset, S. 2015. "Green Space: A New Tool of Public Health Policy? [Les Espaces Verts : Un Nouvel Outil Des Politiques de Santé Publique ?]." *Environnement, Risques et Sante* 14(4):313–20. doi: 10.1684/ers.2015.0795.
65. McCarthy, Mark P., Martin J. Best, and Richard A. Betts. 2010. "Climate Change in Cities Due to Global Warming and Urban Effects." *Geophysical Research Letters* 37(9):1–5. doi: 10.1029/2010GL042845.
66. Meisen P, Quéneudec E, Yuan M, Nara A, et al. 2006. "Overview of Renewable Energy Potential of India." *Modeling Earth Systems and Environment* 2(October):1–20. doi: 10.1016/j.compenvurbsys.2015.03.002.
67. Miller, C. 2008. *Blue-Green Practices: Why They Work and Why They Have Been so Difficult to Implement through Public Policy.*
68. Mohajerani, A., J. Bakaric, and T. Jeffrey-Bailey. 2017. "The Urban Heat Island Effect, Its Causes, and Mitigation, with Reference to the Thermal Properties of Asphalt Concrete." *Journal of Environmental Management* 197:522–38. doi: 10.1016/j.jenvman.2017.03.095.
69. Nabhan, G. P., St Antoine, S., Kellert, S., and Wilson, E. (1993). The loss of floral and faunal story: The extinction of experience. The biophilia hypothesis, 229-250.
70. Najafi, F. T., and H. S. Chaudhry. 2005. "Introducing Practical County and City Management to Undergraduate Students through the Course 'Public Works Engineering and Management Practices.'" Pp. 8845–50 in *ASEE Annual Conference and Exposition, Conference Proceedings*. Portland, OR: American Society for Engineering Education.
71. Newman, G., T. Shi, Z. Yao, D. Li, G. Sansom, K. Kirsch, G. Casillas, and J. Horney. 2020. "Citizen Science-Informed Community Master Planning: Land Use and Built Environment Changes to Increase Flood Resilience and Decrease Contaminant Exposure." *International Journal of Environmental Research and Public Health* 17(2). doi: 10.3390/ijerph17020486.
72. Niesterowicz, Jacek, and Tomasz F. Stepinski, "On Using Landscape Metrics for Landscape Similarity Search." *Ecolical Indicators* 64 May 2016, pp. 20-30. https://doi.org/10.1016/j.ecolind.2015.12.027.
73. Ogutu, A. G., O. P. Kogeda, and M. Lall. 2018. "A Probabilistic Assessment of Location Dependent Failure Trends in South African Water Distribution Networks." Pp. 302–7 in *Lecture Notes in Engineering and Computer Science*. Vol. 2233, edited by A. S. I. C. O. D. C. Korsunsky A.M. Feng D.D. Newswood Limited.

74. Oral, H. V, M. Radinja, A. Rizzo, K. Kearney, T. R. Andersen, P. Krzeminski, G. Buttiglieri, D. Ayral-Cinar, J. Comas, M. Gajewska, M. Hartl, D. C. Finger, J. K. Kazak, H. Mattila, P. Vieira, P. Piro, S. A. Palermo, M. Turco, B. Pirouz, A. Stefanakis, M. Regelsberger, N. Ursino, and P. N. Carvalho. 2021. "Management of Urban Waters with Nature-Based Solutions in Circular Cities—Exemplified through Seven Urban Circularity Challenges." *Water (Switzerland)* 13(23). doi: 10.3390/w13233334.
75. Paul, B., S. S. Bhattacharya, and N. Gogoi. 2021. "Primacy of Ecological Engineering Tools for Combating Eutrophication: An Ecohydrological Assessment Pathway." *Science of the Total Environment* 762. doi: 10.1016/j.scitotenv.2020.143171.
76. Qi, J. D., B. J. He, M. Wang, J. Zhu, and W. C. Fu. 2019. "Do Grey Infrastructures Always Elevate Urban Temperature? No, Utilizing Grey Infrastructures to Mitigate Urban Heat Island Effects." *Sustainable Cities and Society* 46. doi: 10.1016/j.scs.2018.12.020.
77. Ramachandra, T. V, J. Sellers, H. A. Bharath, and B. Setturu. 2019. "Micro Level Analyses of Environmentally Disastrous Urbanization in Bangalore." *Environmental Monitoring and Assessment* 191. doi: 10.1007/s10661-019-7693-8.
78. Rana, Md Masud Parves, and Irina N. Ilina. 2021. "Climate Change and Migration Impacts on Cities: Lessons from Bangladesh." *Environmental Challenges* 5(August):100242. doi: 10.1016/j.envc.2021.100242.
79. Rowe, B. 2016. *Carbon Sequestration and Storage: The Case for Green Roofs in Urban Areas*. Wiley Blackwell.
80. Rubin, V. 2008. *The Roots of the Urban Greening Movement*. University of Pennsylvania Press.
81. Sahni, P., & Kumar, J. (2021). Exploring the relationship of human–nature interaction and mindfulness: a cross-sectional study. *Mental Health, Religion & Culture*, 24(5), 450-462.
82. Santamouris, M., G. Ban-Weiss, P. Osmond, R. Paolini, A. Synnefa, C. Cartalis, A. Muscio, M. Zinzi, T. E. Morakinyo, E. Ng, Z. Tan, H. Takebayashi, D. Sailor, P. Crank, H. Taha, A. L. Pisello, F. Rossi, J. Zhang, and D. Kolokotsa. 2018. "Progress in Urban Greenery Mitigation Science – Assessment Methodologies Advanced Technologies and Impact on Cities." 24(8):638–71. doi: 10.3846/jcem.2018.6604.
83. Scharf, B., and F. Kraus. 2019. "Green Roofs and Greenpass." *Buildings* 9(9). doi: 10.3390/buildings9090205.
84. Shabahang, Sara, Morten Gjerde, Brenda Vale, and Zahra Balador. 2019. *The Problem of Lack of Green Space and Rise in Surface Temperature in the City of Mashhad*. Vol. 131. Springer International Publishing.
85. Shubho, M. T. H., S. R. Islam, B. D. Ayon, and I. Islam. 2015. "An Improved Semiautomatic Segmentation Approach to Land Cover Mapping for Identification of Land Cover Change and Trend." *International Journal of Environmental Science and Technology* 12(8):2593–2602. doi: 10.1007/s13762-014-0649-1.

86. Smith, W. D., S. A. Dunning, S. Brough, N. Ross, and J. Telling. 2020. "GERALDINE (Google Earth Engine SupRaglAciaL Debris INput DEtector): A New Tool for Identifying and Monitoring Supraglacial Landslide Inputs." *Earth Surface Dynamics* 8(4):1053–65. doi: 10.5194/esurf-8-1053-2020.
87. Su, W., Q. Chang, X. Liu, and L. Zhang. 2021. "Cooling Effect of Urban Green and Blue Infrastructure: A Systematic Review of Empirical Evidence." *Shengtai Xuebao* 41(7):2902–17. doi: 10.5846/stxb201903290607.
88. Subiza-Pérez, M., K. Hauru, K. Korpela, A. Haapala, and S. Lehvävirta. 2019. "Perceived Environmental Aesthetic Qualities Scale (PEAQS) – A Self-Report Tool for the Evaluation of Green-Blue Spaces." *Urban Forestry and Urban Greening* 43. doi: 10.1016/j.ufug.2019.126383.
89. Taylor, Lucy, and Dieter F. Hochuli. 2017. "Defining Greenspace: Multiple Uses across Multiple Disciplines." *Landscape and Urban Planning* 158:25–38. doi: 10.1016/j.landurbplan.2016.09.024.
90. Torkianfar, F., H. R. Jafari, and A. Sadeghpour. 2010. "Endangerment Survey of Construction Activities on Shore Line." *Journal of Environmental Studies* 35(52):43–54.
91. Unal Cilek, Muge, and Ahmet Cilek. 2021. "Analyses of Land Surface Temperature (LST) Variability among Local Climate Zones (LCZs) Comparing Landsat-8 and ENVI-Met Model Data." *Sustainable Cities and Society* 69(March):102877. doi: 10.1016/j.scs.2021.102877.
92. Ustaoglu, E., and A. C. Aydınoglu. 2020. "Site Suitability Analysis for Green Space Development of Pendik District (Turkey)." *Urban Forestry and Urban Greening* 47. doi: 10.1016/j.ufug.2019.126542.
93. Vaeztavakoli, Amirafshar, Azadeh Lak, and Tan Yigitcanlar. 2018. "Blue and Green Spaces as Therapeutic Landscapes: Health Effects of Urban Water Canal Areas of Isfahan." *Sustainability (Switzerland)* 10(11). doi: 10.3390/su10114010.
94. Vidhya, K., and R. Shanmugalakshmi. 2020. "Deep Learning Based Big Medical Data Analytic Model for Diabetes Complication Prediction." *Journal of Ambient Intelligence and Humanized Computing* 11(11):5691–5702. doi: 10.1007/s12652-020-01930-2.
95. Villar, R. G., J. L. Pelayo, J. Bantugan, and E. Opiso. 2017. "Algorithm for Modeling Agricultural Land Cover Classification and Land Surface Temperature." Pp. 112–23 in *GISTAM 2017 - Proceedings of the 3rd International Conference on Geographical Information Systems Theory, Applications and Management*, edited by R. J. G. Ragia L. Laurini R. SciTePress.
96. Wang, Y., Y. Li, Y. Xue, A. Martilli, J. Shen, and P. W. Chan. 2020. "City-Scale Morphological Influence on Diurnal Urban Air Temperature." 169. doi: 10.1016/j.buildenv.2019.106527.
97. Weng, Qihao, Umamaheshwaran Rajasekar, and Xuefei Hu. 2011. "Modeling Urban Heat Islands and Their Relationship with Impervious Surface and Vegetation Abundance by Using ASTER Images." *IEEE Transactions on Geoscience and Remote Sensing* 49(10):4080–89. doi: 10.1109/TGRS.2011.2128874.

98. Wilczyńska, A., I. Myszka, S. Bell, M. Slapińska, N. Janatian, and A. Schwerk. 2021. "Exploring the Spatial Potential of Neglected or Unmanaged Blue Spaces in the City of Warsaw, Poland." *Urban Forestry and Urban Greening* 64. doi: 10.1016/j.ufug.2021.127252.
99. Workie, T. G., and H. J. Debella. 2018. "Climate Change and Its Effects on Vegetation Phenology across Ecoregions of Ethiopia." *Global Ecology and Conservation* 13. doi: 10.1016/j.gecco.2017.e00366.
100. Wu, C., J. Li, C. Wang, C. Song, Y. Chen, M. Finka, and D. La Rosa. 2019. "Understanding the Relationship between Urban Blue Infrastructure and Land Surface Temperature." *Science of the Total Environment* 694. doi: 10.1016/j.scitotenv.2019.133742.
101. Xiao, F., and J. Q. Wang. 2019. "Multistage Decision Support Framework for Sites Selection of Solar Power Plants with Probabilistic Linguistic Information." *Journal of Cleaner Production* 230:1396–1409. doi: 10.1016/j.jclepro.2019.05.138.
102. Yang, F., S. S. Y. Lau, and F. Qian. 2010. "Summertime Heat Island Intensities in Three High-Rise Housing Quarters in Inner-City Shanghai China: Building Layout, Density and Greenery." 45(1):115–34. doi: 10.1016/j.buildenv.2009.05.010.
103. Yoo, D., J. Jung, W. Jeong, and S. Han. 2021. "Metadynamics Sampling in Atomic Environment Space for Collecting Training Data for Machine Learning Potentials." *Npj Computational Materials* 7(1). doi: 10.1038/s41524-021-00595-5.
104. Yu, Z., L. Di, J. Tang, C. Zhang, L. Lin, E. G. Yu, M. S. Rahman, J. Gaigalas, and Z. Sun. 2018. "Land Use and Land Cover Classification for Bangladesh 2005 on Google Earth Engine." in *2018 7th International Conference on Agro-Geoinformatics, Agro-Geoinformatics 2018*. Institute of Electrical and Electronics Engineers Inc.
105. Zari, M. P. 2019. "Devising Urban Biodiversity Habitat Provision Goals: Ecosystem Services Analysis." *Forests* 10(5). doi: 10.3390/f10050391.
106. Zhang, C., C. Wang, C. Chen, L. Tao, J. Jin, Z. Wang, and B. Jia. 2022. "Effects of Tree Canopy on Psychological Distress: A Repeated Cross-Sectional Study before and during the COVID-19 Epidemic." *Environmental Research* 203. doi: 10.1016/j.envres.2021.111795.
107. Zhang, D. D., and L. Zhang. 2020. "Land Cover Change in the Central Region of the Lower Yangtze River Based on Landsat Imagery and the Google Earth Engine: A Case Study in Nanjing, China." *Sensors (Switzerland)* 20(7). doi: 10.3390/s20072091.
108. Zhang, Y., M. A. Tarrant, and G. T. Green. 2008. "The Importance of Differentiating Urban and Rural Phenomena in Examining the Unequal Distribution of Locally Desirable Land." *Journal of Environmental Management* 88(4):1314–19. doi: 10.1016/j.jenvman.2007.07.008.

109. Zhao, T. F., and K. F. Fong. 2017. "Characterization of Different Heat Mitigation Strategies in Landscape to Fight against Heat Island and Improve Thermal Comfort in Hot-Humid Climate (Part II): Evaluation and Characterization." *Sustainable Cities and Society* 35:841–50. doi: 10.1016/j.scs.2017.05.006.
110. Zurqani, H. A., C. J. Post, E. A. Mikhailova, M. P. Cope, J. S. Allen, and B. A. Lytle. 2020. "Evaluating the Integrity of Forested Riparian Buffers over a Large Area Using LiDAR Data and Google Earth Engine." *Scientific Reports* 10(1). doi: 10.1038/s41598-020-69743-z.

12
Synthesis and Future Perspective

Deepak Kumar[1,2]

[1]Center of Excellence in Weather & Climate Analytics, Atmospheric Sciences Research Center (ASRC), University at Albany (UAlbany), State University of New York (SUNY), Albany, New York, USA
[2]Amity Institute of Geoinformatics & Remote Sensing (AIGIRS), Amity University Uttar Pradesh (AUUP), Gautam Buddha Nagar, Uttar Pradesh, India

Abstract

The lack of available space for renewable energy installations is a major obstacle in urban areas. The equilibrium between city energy demand and renewable energy density serves as the basis for our analytical methodology for decarbonized urban environments. Waves of innovation in the energy sector could come from a variety of sources in the future. These sources include solar electricity generated in space, nuclear power facilities that can be disassembled and reused, and deep geothermal systems. If these techniques are put into practice, greenhouse gas emissions and the need for clean water and air may decrease. To meet the energy needs of modern cities while also lowering their carbon footprint, widespread adoption of renewable energy sources is essential. Improvements in efficiency, usability, cost-effectiveness, accessibility, and sustainability are all on the horizon for renewable energy sources.

Keywords: Energy source, energy demand, energy policy, future energy

12.1 Introduction

Due to the global energy problem and climate change, industrialized and developing countries must innovate in energy and consume responsibly.

Email: deepakdeo2003@gmail.com

Deepak Kumar (ed.) Urban Energy Systems: Modeling and Simulation for Smart Cities, (193–204)
© 2023 Scrivener Publishing LLC

Rich nations divorced economic growth from energy use to reduce resource use and environmental implications. Waste heat was used to generate power [1, 2]. Decarbonizing and improving energy efficiency are crucial. Despite fossil fuels' continued relevance in urban energy generation, renewables remain the only feasible option moving ahead [3, 4]. In cities, cogeneration and district heating may be crucial [5, 6]. Renewable energy becomes "energy-important" as cities grow. Changing the energy source is just the beginning; the transition will be successful if the new source is affordable, durable, and growth-friendly [7, 8]. Solar power doesn't need water, alleviating water demand and scarcity worries. Recent improvements in solar technology (especially concentrated and photovoltaic solar power) have decreased implementation costs, making them competitive with fossil fuel-based power generation in mid- and high-latitude areas [9, 10]. Solar panels and photovoltaics can be put on roofs with minimum aesthetic impact, making them perfect for urban locations. Future improvements in on-site renewable energy production may allow zero-emission buildings and energy-efficient eco-cities. Each day, new technology makes cities greener and more self-sufficient. Wind, sun, and rain harvesters are being designed for tall structures to increase energy generation. This decreases urban wind turbine risks. Advances in technology may boost eco-cities [11, 12]. Urban neighbourhoods have water-efficient fixtures, double-paned windows, a southerly orientation for passive heating, solar photovoltaic roofs and walls, and power-generating stations. Incompatibility between supply and demand and their integration into the energy system may hinder the adoption of renewable energies in metropolitan areas. With smart networks, electricity distribution may be better regulated [13–15].

Implementing such methods in an urban setting improves energy security and reliability, reduces transmission costs, uses existing infrastructure, and reduces land requirements. International or regional certification systems that meet high standards without ignoring local cultures would assist the sector. Long-term requirements have made the energy industry stable and profitable [16]. The government promotes tiny residential solar water heating systems among low-income households or social housing complexes to cut home energy costs, give employment opportunities, and keep up with renewable energy demand. Low-interest financing and solar district heating are other options. Low-carbon, resilient, and liveable cities are sustainable [17, 18].

Climate change, extreme weather, and natural and man-made disasters threaten urban energy infrastructure. A city's energy infrastructure affects health and happiness. High-energy businesses like transport, food production, and water quality affect health, economic competitiveness, cultural appeal, and social, gender, and racial equality. Decentralized, non-fossil fuel energy may help cities accomplish sustainability goals. Renewable energy increases self-sufficiency and resilience. Greater power densities of renewable energy technologies, enhanced infrastructure that can handle large integrated energy systems, and better urban energy efficiency, especially in buildings, may assist in solving these problems [19–22].

Recent technological advances may make cities greener. Environmental, economic, and social sustainability may all benefit from city-integrated renewable energy, which generates energy where it's required. Distributed energy systems offer four benefits: carbon-neutral products and services, fewer costly network improvements, decentralized power generation and communication, and social harmony.

12.2 Synthesis of the Research

Most cities comprehend local and global sustainability. It requires perceiving a city as a complex and dynamic ecosystem, open system, or cluster of systems where natural resources are modified to fit urban needs. The techno-economic analysis finds: If the proposed solar power plant is put on the roof of a normal home, the homeowner may create excess electricity, sell it to the grid, and reduce annual carbon dioxide emissions. Better information and forecasting of power demand and generation, especially in urban areas with dense networks, are vital for implementing associated policies in many nations. Models of urban energy systems are vital for evaluating improved designs, new policies, and related technology, as seen here. In recent decades, models have proliferated across formulations, applications, times, and countries. No work reveals the full scope of the activity in this area, and no resource helps understand and make sense of the writing. In this study, we conduct a thorough literature review on urban energy system modelling to fill the knowledge gap. Underdeveloped cities are moving from biomass to fossil fuels and electricity. Second-generation energies are easy to use and have fewer environmental implications. The fuel change coincides with urban growth.

The claims are backed by empirical evidence. The survey's fuel shift findings are consistent with those in other emerging-market cities.

Due to differences in access to petroleum and machinery, government policies favour the wealthy. Energy models are vital for implementing energy transitions because they accurately portray energy-related difficulties. A recent trend analysis of energy system models found more emphasis on open access, cross-sectoral modelling, and temporal detail. This was done to plan for future scenarios with several varied RE sources. "Major challenges remain in expressing high-resolution energy demand across all sectors," as well as "openness and accessibility," "how tools are coupled," and "tool developers' and policy/decision-makers involvement."

To plan and carry out such intervention drives, there must be proof of niches that would directly profit from such a strategy, and all potential power supply choices must be examined and analyzed. Both grid-connected and off-grid communities need evaluations of green power assets and energy overviews. Sustainable power systems will be established and their yearly power supply evaluated based on energy studies and other optional information. While considering the investigation's systems, we will also analyze the customer's willingness and ability to pay for green power. Considering the above studies on renewable energy, one can see the importance of generating electricity from renewable energy in the next years and how to overcome low efficacy and partial shading sway. This research can be used to construct a model that accounts for future energy consumption and load profiles. The findings will aid places that lack solar energy knowledge and can't produce sustainable electricity. Local community and younger pupils will be reached through awareness-raising activities to promote the shift to renewable energy sources and encourage energy saving. Renewable energy sources must be widely adopted to meet cities' expanding energy needs and minimize their carbon footprint (ibid). As technology improves, renewable energy sources will be more efficient, usable, cost-effective, accessible, and sustainable.

Archetypes are based on building physics parameters such as floor space, window-to-wall ration (WWR), kind of housing, building use, climatic zone, U-Value, etc. MIT Design Advisor's rapid building energy modelling tool estimates total annual energy consumption for cooling and lighting load. Simulated typology findings were scaled to the entire case area's buildings, and urban energy consumption and performance maps were created. Pilot research demonstrates district-scale urban energy measurement and rooftop solar photovoltaic potential. Using the same procedure, the entire city's energy potential can be analysed. Architects, urban planners, and decision-makers can use these tools to quickly assess and quantify energy generation potentials. Also, the urban rooftop solar PV potential of the case region was analysed. The work also featured 3D urban models and 3D GIS for visualization and decision-making. Considering the retrieved building profile and context, 3D models may be more accurate. 3D data collecting and modelling need to be simplified.

By increasing the capacity of nuclear power plants and other renewable energy sources, as well as making full use of hydropower, the demand for coal can be reduced, allowing for the construction of more hybrid scenarios. Regulations on the release of carbon dioxide can be utilized as constraints to investigate trends in greater depth.

India is perfect for solar electricity. Solar electricity will benefit millions of Indians, especially the impoverished. It can support growth in all sectors of development and create thousands of jobs in the solar sector, helping India become the "India of the future." Rooftop solar panels remain a feasible sector for solar energy use through programs and efforts that may help India break new ground in solar power generation and position the country to become a global leader in green energy. India's tropical location, an enormous market, legislative incentives, and educational and research organizations make it an energy powerhouse. Insufficient equipment maintenance and difficult-to-negotiate fees make government-provided energy expensive and unreliable. India's National Solar Mission promotes renewable energy. The inability to maintain solar plants may limit program feasibility.

Climate change and sustainable development require clean energy. Globally, renewable energy has low/zero carbon emissions. Weather issues affect the hybrid renewable sector, but a microgrid can help. Microgrid energy management reduces peaks and shifts. Microgrids manage energy using demand and output forecasts. Distributed generation (DG) and energy storage are vital to micro/smart grids. This work summarizes prominently published research papers essential for future study since we need to learn about energy management systems in microgrids using AI, Fuzzy logic, and other methodologies. The work analysed clean energy conversion methods for energy management systems. It explored intelligent scheduling models and analysis models.

Indian cities, like Bengaluru, struggle to manage waste. These difficulties, either solved or prioritized, are the initial step in waste-to-energy conversion. The conclusions were based on successful and failing Indian plants and conversations with specialists. The study also found that waste management must start with households. The understanding spans four themes: environmental, technological, social, and economic. Environmentally, putting trash in landfills is unsustainable. Sending unsegregated waste to waste-to-energy facilities causes extra pollution due to low waste quality and affects plant efficiency. On the technology level, ULBs lack state and federal funds to install city-level plants. The city must subsidize the energy plant's electricity and install garbage under the PPP approach. It's uneconomical and inefficient. Land, water, and air pollution from non-segregated waste-to-energy plants harm citizens. A basic economic difficulty is ULBs' failure to compensate and incentivise city rubbish pickers. Therefore, waste pickers turn to unscrupulous tactics to earn, which lowers the quality of gathered waste for processing. Through the suggestion at the city, neighbourhood and household levels, the authors hope to give direction to abate the challenges of waste management and contribute to making our cities more sustainable.

Energy waste is a potential concept in many wealthy countries. China, Japan, Denmark, etc., have started extracting value-added products and energy from urban solid garbage. Developing countries lag due to the aforesaid factors. Developing countries have tremendous opportunities with waste-to-energy, including employment, household income growth, better health, and sanitation, etc. Legacy garbage in outengineered landfills impedes growth and progress. Thus, research should focus on discovering better ways for treating legacy garbage since it may be reduced at source by educating the public.

Change detection analysis using multi-temporal remote sensing pictures and GIS helped planners and administrators understand land use patterns. This region can have sustainable land management and city planning. Gandhinagar has expanded in recent decades. Remote sensing, GIS, and entropy captured urban growth's spatial dynamics. This study shows that built-up areas around the city centre are expanding outward. Study results can be used for sustainable practices, such as solar energy expansion. The study shows development along the diverting and straight roads.

Human activity causes urban microclimates to be warmer than their surroundings and what is expected. Cities are called "heat islands." The city centre is several degrees warmer than the countryside when conditions are calm. Urban microclimates are warmer due to human involvement. Even on a moderate day, temperatures increase in a city's centre and fall in the suburbs and countryside. This trend has several causes. The retained heat makes cities uninhabitable. Dark tarmac and concrete absorb a lot of heat due to their high thermal capacity. The daytime heat is discharged in the evening, raising the temperature. The city's manufacturing and rising car use contribute to haze and a pollution dome. This pollution dome allows short-wave insolation but traps longer-wave outgoing terrestrial radiation, increasing heat. Natural interactions provide health and well-being. Early childhood is the time to form these bonds with nature. This will lead to health benefits like blood pressure regulation, psychological benefits like resilience, and food security by analyzing food growth. Healthy eating habits involve building a pleasant relationship with food. Increasing this relationship requires combining environmental and food-related efforts. These activities create social support, trust, and cohesion, reducing food insecurity. These programs give farmers new markets and strengthen the food chain.

12.3 Future Urban Energy Policies, and Initiatives

Energy is crucial to progress no matter where it comes from. Transport, manufacturing, commerce, construction, infrastructure, water supply, and agriculture all rely on it. Grid flexibility must be significantly improved in the future power system to support substantial proportions of variable renewables to solve global climate change. An integrated methodology for modelling the growing complexity and interdependence of energy systems is necessary to quantify the sector coupling opportunities at the city level. Specific to the significance of the buildings sector in the pursuit of net-zero cities and communities, the paper also presents a modelling tool to allow quantitative evaluation of sector coupling prospects in each urban stakeholder to realize that cities have tremendous potential to accelerate and benefit from the race to net zero and to aid the decision-making process inside cities transitioning towards a net zero future. The urban stakeholders realize that cities have tremendous potential to accelerate and benefit from the race to net zero and to aid the decision-making process inside cities transitioning towards a net zero future.

The potential for cross-sector synergies to be realized with sector-coupling technology and the availability of trade-off alternatives can be evaluated about optimization goals in various contexts. Until battery storage technologies can compete with fossil fuel power plants, non-RE electricity generation will be needed. Due to dropping lithium-ion battery prices, this will be possible soon. India's power sector is the largest contributor to energy-related greenhouse gas emissions. According to the study, "low-quality coal used in inefficient power plants has led to urban air pollution, compounding other environmental problems." Several policy actions, including the green hydrogen policy, offshore wind policy, promotion of electric vehicles, and green day-ahead market, indicate the government's determination. India's energy sources are coal and oil. Replacing renewable energy is expensive. The administration has shown its commitment by promoting green hydrogen, offshore wind, electric vehicles, a green day-ahead market, and open access to green energy. India's energy sources are coal and oil. Substitute green energy.

Cities are crucial to the global energy transition because they utilize most of the world's primary energy for building construction and operation, logistics, urban mobility, industry, and other uses. Cities produce more than half of all GHGs. The urban population will grow by 2 billion by 2050, largely in emerging Asia and Africa, highlighting the importance of renewables in meeting rising energy demand and decreasing greenhouse gas emissions and other repercussions of fossil fuel consumption.

For example, once a renewable energy generation system is set up, the ongoing expenses are often quite low because the energy source is both cheap and abundant. To successfully convert the local climate and geography into usable energy, communities must consider both. Renewable energy sources, such as the sun at night or wind during doldrums, should be integrated to combat intermittency. Waste and heat recovery technologies offer another potential source of relief by filling in supply shortages.

Another option for dealing with unreliable power supply is the implementation of smart grids or electric grids that balance supply and demand. Reduced energy loss during transmission and distribution is another benefit of these systems. It is possible to program machine shifts to operate only during times of the day when there is sufficient electricity supply (for example, washing machines do not need to run at a specific time, so they can turn on automatically while the customer is asleep, or at work).

Cities are well-positioned to encourage renewable energy in buildings, transportation, industry, and power. Municipal authorities oversee urban zoning, construction permits, and solar legislation. They can levy local taxes, provide low-interest loans, and issue municipal green bonds. Municipalities own or operate energy-generating facilities and urban infrastructure. Cities have the most obvious energy changes. Large cities have the legislative frameworks and infrastructure to scale up renewables and meet emission reduction targets. Small and medium-sized cities (less than 1 million people) lack sufficient funding and legislative support to develop on this path. They are home to 2.4 billion people, or 59% of the world's urban population, and are increasing faster than any other urban group. The widespread usage of fossil fuels has several negative consequences. Rising greenhouse gas emissions and pollution from burning fuels aren't the only problems with carbon-based energy generation; extraction methods can also leave harmful chemicals behind, and mishaps during production and distribution are all too frequent. In addition, economies are vulnerable to supply shocks or price spikes in fossil fuels because of their current reliance on these commodities. Unfortunately, fossil fuels are also frequently utilized as a tool of political pressure and are a source of conflict in the region. Additionally, the depletion of fossil fuel reserves is an imminent possibility.

12.4 The Challenge Ahead

Seventy-five percent of the world's primary energy is used in urban areas, thus they need a constant supply. Energy distribution must become sustainable, more egalitarian, and fair to support universal development while also reducing urban areas' ecological footprint. The yearly demand growth for energy is roughly 7% in developing countries, whereas the supply is relatively consistent despite the growing population, fast urbanization, and expanding economic development. Thus, there is a discrepancy between supply and demand, resulting in regular power rationing in urban areas.

12.5 Strategies for Improvement

Sustainable urban energy systems require low-carbon technology on the supply side, efficient distribution infrastructure, and reduced consumption on the part of end users. Therefore, cities need to switch from using unsustainable fossil fuels to generate electricity to using renewable energy sources, not just because of the impending depletion of resources but also to reduce the negative externalities that result from using fossil fuels, such as pollution and greenhouse gas emissions. Also, we need to change our consumption habits and implement energy-saving measures to lower our overall energy usage.

To create urban energy systems that are sustainable and to decrease the use of unsustainable technologies and practices, cities need to establish strong policies and standards. Legislation governing energy use and consumption is necessary, but governments should also implement incentive mechanisms to spur exploration, innovation, and, most crucially, the widespread adoption of cleaner, more efficient technology. The private sector operates the majority of the world's energy systems, so there must be strong collaboration and mutual understanding between this sector and the authorities in charge of monitoring it if short-term commercial interests are not to overshadow long-term environmental concerns and sustainable development opportunities.

Improved cost-effectiveness is possible by using energy-saving heating, cooling, insulation, lighting, and water distribution systems in newly constructed or renovated structures. Similarly, power from the grid can be complemented by on-site renewable energy sources like solar panels on a roof. Better energy balance can also be achieved using recycled, reused, or low-energy building materials. Attractive public transportation systems, an increased percentage of non-motorized transport in constructing specific infrastructure (such as bike lanes and sidewalks) and optimizing the delivery of commodities are all necessary for communities to reduce their reliance on fossil fuels for transportation needs (for instance by promoting the use of rail for cargo transport). The energy needs for providing food and water to people are substantial. Governments should encourage local and international collaboration to help local enterprises improve their knowledge, skills, and market reach. Developing countries should pursue private-public partnerships to develop their energy systems due to high costs. For each city to adapt to its unique particularities, the authorities must develop decentralized energy systems and infrastructure and be allowed to have specific regulations and tax systems to promote sustainable energy or restrict polluting inefficient technology and consumption practices.

Finally, as energy is critical to earning money, there is a pressing need for more equitable energy distribution to promote universal development, particularly for the urban poor. Although RETs such as wind, water, solar, and geothermal are becoming more accessible and are meeting the energy needs of some neighbourhoods if not entire cities in some locations, unreliable supply and high upfront costs are preventing them from becoming widespread. However, the long-term benefits, both environmentally and economically, will surpass the initial difficulties.

Consumers (individuals, organizations, and institutions) are ultimately responsible for making significant changes by reducing their consumption. Less consumption means less generation capacity is required. Saving energy is another area where the technology may help out. To adapt to shifts in electricity supply and pricing, smart grids can work in tandem with smart appliances or even an entire smart home or building. Smart meters enable homes, businesses, and factories to schedule the use of specific appliances during times of high electricity availability. A washing machine, for instance, can be programmed to only begin running when the price of electricity drops below a specific point or when there is sufficient power in the grid. Green or low-energy building concepts and passive design principles can significantly reduce a building's energy consumption.

References

1. M. H. Babikir *et al.*, "Simplified Modeling and Simulation of Electricity Production from a Dish/Stirling System," *Int. J. Photoenergy*, vol. 2020, 2020, doi: 10.1155/2020/7398496.
2. M. Irfan *et al.*, "Assessing the energy dynamics of Pakistan: Prospects of biomass energy," *Energy Reports*, vol. 6, pp. 80–93, 2020, doi: 10.1016/j.egyr.2019.11.161.
3. C. Gschwendtner, C. Knoeri, and A. Stephan, "The impact of plug-in behavior on the spatial–temporal flexibility of electric vehicle charging load," *Sustain. Cities Soc.*, vol. 88, 2023, doi: 10.1016/j.scs.2022.104263.
4. C. Wang *et al.*, "Thermal environment and thermal comfort in metro systems: A case study in severe cold region of China," *Build. Environ.*, vol. 227, 2023, doi: 10.1016/j.buildenv.2022.109758.
5. E. Guidetti and M. Ferrara, "Embodied energy in existing buildings as a tool for sustainable intervention on urban heritage," *Sustain. Cities Soc.*, vol. 88, 2023, doi: 10.1016/j.scs.2022.104284.
6. S. Garshasbi *et al.*, "On the energy impact of cool roofs in Australia," *Energy Build.*, vol. 278, 2023, doi: 10.1016/j.enbuild.2022.112577.
7. D. Pan, Y. Bai, M. Chang, X. Wang, and W. Wang, "The technical and economic potential of urban rooftop photovoltaic systems for power generation in Guangzhou, China," *Energy Build.*, vol. 277, 2022, doi: 10.1016/j.enbuild.2022.112591.
8. J. Fu, S. Hu, X. He, S. Managi, and D. Yan, "Identifying residential building occupancy profiles with demographic characteristics: using a national time use survey data," *Energy Build.*, vol. 277, 2022, doi: 10.1016/j.enbuild.2022.112560.
9. X. Li *et al.*, "Application of effective water-energy management based on digital twins technology in sustainable cities construction," *Sustain. Cities Soc.*, vol. 87, 2022, doi: 10.1016/j.scs.2022.104241.
10. P. Mehta and V. Tiefenbeck, "Solar PV sharing in urban energy communities: Impact of community configurations on profitability, autonomy and the electric grid," *Sustain. Cities Soc.*, vol. 87, 2022, doi: 10.1016/j.scs.2022.104178.
11. P. J. Vergragt, L. Dendler, M. de Jong, and K. Matus, "Transitions to sustainable consumption and production in cities," *J. Clean. Prod.*, vol. 134, no. Part A, pp. 1–12, 2016, doi: 10.1016/j.jclepro.2016.05.050.
12. I. Andresen *et al.*, "Design and performance predictions of plus energy neighbourhoods – Case studies of demonstration projects in four different European climates," *Energy Build.*, vol. 274, 2022, doi: 10.1016/j.enbuild.2022.112447.
13. M. J. Marquez-Ballesteros, L. Mora-López, P. Lloret-Gallego, A. Sumper, and M. Sidrach-de-Cardona, "Measuring urban energy sustainability and its application to two Spanish cities: Malaga and Barcelona," *Sustain. Cities Soc.*, vol. 45, pp. 335–347, 2019, doi: 10.1016/j.scs.2018.10.044.

14. M. Engelken, B. Römer, M. Drescher, and I. Welpe, "Transforming the energy system: Why municipalities strive for energy self-sufficiency," *Energy Policy*, vol. 98, pp. 365–377, 2016, doi: 10.1016/j.enpol.2016.07.049.
15. H. Z. Al Garni and A. Awasthi, "Solar PV power plant site selection using a GIS-AHP based approach with application in Saudi Arabia," *Appl. Energy*, vol. 206, pp. 1225–1240, 2017, doi: 10.1016/j.apenergy.2017.10.024.
16. F. Boudali Errebai, D. Strebel, J. Carmeliet, and D. Derome, "Impact of urban heat island on cooling energy demand for residential building in Montreal using meteorological simulations and weather station observations," *Energy Build.*, vol. 273, 2022, doi: 10.1016/j.enbuild.2022.112410.
17. M. Angelidou *et al.*, "Governing the smart city: a review of the literature on smart urban governance," *Cities*, vol. 82, no. 2, pp. 95–106, 2016, doi: 10.1177/0974928416637930.
18. J. Blair, C. Roldan, S. Ghosh, and S.-H. Yung, "Greening Rail Infrastructure for Carbon Benefits," in *Procedia Engineering*, 2017, vol. 180, pp. 1716–1724, doi: 10.1016/j.proeng.2017.04.334.
19. T. Long *et al.*, "Constrained crystals deep convolutional generative adversarial network for the inverse design of crystal structures," *NPJ Comput. Mater.*, vol. 7, no. 1, 2021, doi: 10.1038/s41524-021-00526-4.
20. Y. Shi, D. Feng, S. Yu, C. Fang, H. Li, and Y. Zhou, "The projection of electric vehicle population growth considering scrappage and technology competition: A case study in Shanghai," *J. Clean. Prod.*, vol. 365, 2022, doi: 10.1016/j.jclepro.2022.132673.
21. J. Byrne, M. Mouritz, M. Taylor, and J. K. Breadsell, "East village at Knutsford: A case study in sustainable urbanism," *Sustain.*, vol. 12, no. 16, 2020, doi: 10.3390/SU12166296.
22. K. Petrauskiene *et al.*, "Situation analysis of policies for electric mobility development: Experience from five european regions," *Sustain.*, vol. 12, no. 7, 2020, doi: 10.3390/su12072935.

About the Editor

Deepak Kumar, PhD, is a research scientist at the Center of Excellence in Weather and Climate Analytics, Atmospheric Science Research Center, State University of New York, University at Albany, New York, USA, with over ten years of experience. He has published 3 books. He has also published over 45 and reviewed over 190 research articles in various scientific and scholarly journals, and he has completed two research projects sponsored by the Science and Engineering Research Board, Department of Science and Technology, Government of India.

Index

Analytical hierarchy process (AHP), 196
Applications, 12, 35, 73
ArcGIS, 53

Biodegradable, 120, 145

Classes, 161, 163
Climate change, 194, 197
Climate resilience, 172, 197
CO_2, 113
Communities, 6, 9, 198, 200
Component, 113, 114, 142, 173, 178

Decarbonizing, 178
Disaster, 19, 112, 195

Economic performance, 8, 28, 76
Effluent, 152
Emerging market, 195
Energy consumption, 173, 174, 196, 202
Energy demand, 69, 83, 84, 111
Energy policy, 72, 75, 82, 193
Energy resources, 8, 21, 32, 94
Energy source, 32, 33, 44, 48, 49, 61
Environmental pollution, 122

Future energy, 82, 193, 196

Geographic information system (GIS), 11, 49, 53, 179
Geographical information system (GIS), 53, 58, 62
GIS techniques, 158, 168
GIS, 11

GPS, 24, 31
Greater population, 4

Health, 99, 122, 124, 133

Land cover, 155, 160, 163
Land use, 53, 74, 91, 155
Landsat, 156, 159
LULC statistics, 192
LULC, 155, 156, 160

Metropolitan households, 2

Petroleum and machinery, 194
Planning, 166, 169, 171

Remote sensing and GIS, 165
Remote sensing, 196
Renewable energy demand, 192
Renewable energy, 192, 193, 195, 198

Satellite images, 158, 159, 161
Slope, 11
Solar energy expansion, 167, 196
Sustainable and environmentally, 3

Temperature, 2, 31, 38, 50, 196
Transmission costs, 192

UHI, 170
Urban heat island, 50, 172
User's accuracy, 162
USGS, 158

Wastewater, 121, 126, 151, 177

Also of Interest

Check out these other related titles from Scrivener Publishing

ENERGY STORAGE TECHNOLOGIES IN GRID MODERNIZATION, Edited by Sandeep Dhundhara, Yajvender Pal Verma, and Ashwani Kumar, ISBN: 9781119872115. Written and edited by a team of experts, this exciting new volume discusses the various types of energy storage technologies, the applications of energy storage systems, the performance improvement of modern power systems, their role in the real-time operation of power markets, and the operational issues of modern power systems, including renewable-based generating sources.

SMART GRIDS FOR SMART CITIES VOLUME 1, Edited by O.V. Gnana Swathika, K. Karthikeyan, and Sanjeevikumar Padmanaban, ISBN: 9781119872078. Written and edited by a team of experts in the field, this first volume in a two-volume set focuses on an interdisciplinary perspective on the financial, environmental, and other benefits of smart grid technologies and solutions for smart cities.

SMART GRIDS FOR SMART CITIES VOLUME 2: Real-Time Applications in Smart Cities, Edited by O.V. Gnana Swathika, K. Karthikeyan, and Sanjeevikumar Padmanaban, ISBN: 9781394215874. Written and edited by a team of experts in the field, this second volume in a two-volume set focuses on an interdisciplinary perspective on the financial, environmental, and other benefits of smart grid technologies and solutions for smart cities.

SMART GRIDS AND INTERNET OF THINGS, Edited by Sanjeevikumar Padmanaban, Jens Bo Holm-Nielsen, Rajesh Kumar Dhanaraj, Malathy Sathyamoorthy, and Balamurugan Balusamy, ISBN: 9781119812449. Written and edited by a team of international professionals, this groundbreaking new volume covers the latest technologies in automation, tracking, energy distribution and consumption of Internet of Things (IoT) devices with smart grids.

SMART GRIDS AND GREEN ENERGY SYSTEMS, Edited by A. Chitra, V. Indragandhi and W. Razia Sultana, ISBN: 9781119872030. Presenting the concepts and advances of smart grids within the context of "green" energy systems, this volume, written and edited by a global team of experts, goes into the practical applications that can be utilized across multiple disciplines and industries, for both the engineer and the student.

SMART GRIDS AND MICROGRIDS: Concepts and Applications, Edited by P. Prajof, S. Mohan Krishna, J. L. Febin Daya, Umashankar Subramaniam, and P. V. Brijesh, ISBN: 9781119760559. Written and edited by a team of experts in the field, this is the most comprehensive and up to date study of smart grids and microgrids for engineers, scientists, students, and other professionals.

MICROGRID TECHNOLOGIES, Edited by C. Sharmeela, P. Sivaraman, P. Sanjeevikumar, and Jens Bo Holm-Nielsen, ISBN 9781119710790. Covering the concepts and fundamentals of microgrid technologies, this volume, written and edited by a global team of experts, also goes into the practical applications that can be utilized across multiple industries, for both the engineer and the student.

INTEGRATION OF RENEWABLE ENERGY SOURCES WITH SMART GRIDS, Edited by A. Mahaboob Subahani, M. Kathiresh and G. R. Kanagachidambaresan, ISBN: 9781119750420. Provides comprehensive coverage of renewable energy and its integration with smart grid technologies.

Encyclopedia of Renewable Energy, by James G. Speight, ISBN 9781119363675. Written by a highly respected engineer and prolific author in the energy sector, this is the single most comprehensive, thorough, and up to date reference work on renewable energy.

Green Energy: Solar Energy, Photovoltaics, and Smart Cities, edited by Suman Lata Tripathi and Sanjeevikumar Padmanaban, ISBN 9781119760764. Covering the concepts and fundamentals of green energy, this volume, written and edited by a global team of experts, also goes into the practical applications that can be utilized across multiple industries, for both the engineer and the student.

Energy Storage, edited by Umakanta Sahoo, ISBN 9781119555513. Written and edited by a team of well-known and respected experts in the field, this new volume on energy storage presents the state-of-the-art developments and challenges in the field of renewable energy systems for sustainability and scalability for engineers, researchers, academicians, industry.

Energy Storage 2nd Edition, by Ralph Zito and Haleh Ardebili, ISBN 9781119083597. A revision of the groundbreaking study of methods for storing energy on a massive scale to be used in wind, solar, and other renewable energy systems.

Hybrid Renewable Energy Systems, edited by Umakanta Sahoo, ISBN 9781119555575. Edited and written by some of the world's top experts in renewable energy, this is the most comprehensive and in-depth volume on hybrid renewable energy systems available, a must-have for any engineer, scientist, or student.

Progress in Solar Energy Technology and Applications, edited by Umakanta Sahoo, ISBN 9781119555605. This first volume in the new groundbreaking series, Advances in Renewable Energy, covers the latest concepts, trends, techniques, processes, and materials in solar energy, focusing on the state-of-the-art for the field and written by a group of world-renowned experts.

A Polygeneration Process Concept for Hybrid Solar and Biomass Power Plants: Simulation, Modeling, and Optimization, by Umakanta Sahoo, ISBN 9781119536093. This is the most comprehensive and in-depth study of the theory and practical applications of a new and groundbreaking method for the energy industry to "go green" with renewable and alternative energy sources.

DESIGN AND DEVELOPMENT OF EFFICIENT ENERGY SYSTEMS, edited by Suman Lata Tripathi, Dushyant Kumar Singh, Sanjeevikumar Padmanaban, and P. Raja, ISBN 9781119761631. Covering the concepts and fundamentals of efficient energy systems, this volume, written and edited by a global team of experts, also goes into the practical applications that can be utilized across multiple industries, for both the engineer and the student.

INTELLIGENT RENEWABLE ENERGY SYSTEMS: Integrating Artificial Intelligence Techniques and Optimization Algorithms, edited by Neeraj Priyadarshi, Akash Kumar Bhoi, Sanjeevikumar Padmanaban, S. Balamurugan, and Jens Bo Holm-Nielsen, ISBN 9781119786276. This collection of papers on artificial intelligence and other methods for improving renewable energy systems, written by industry experts, is a reflection of the state of the art, a must-have for engineers, maintenance personnel, students, and anyone else wanting to stay abreast with current energy systems concepts and technology.

SMART CHARGING SOLUTIONS FOR HYBRID AND ELECTRIC VEHICLES, edited by Sulabh Sachan, Sanjeevikumar Padmanaban, and Sanchari Deb, ISBN 9781119768951. Written and edited by a team of experts in the field, this is the most comprehensive and up to date study of smart charging solutions for hybrid and electric vehicles for engineers, scientists, students, and other professionals.

Printed and bound by CPI Group (UK) Ltd, Croydon, CR0 4YY
20/08/2023

08102133-0001